T0135546

On Representability
of ∗-Regular and Regular Involutive Rings
in Endomorphism Rings of Vector Spaces

Vom Fachbereich Mathematik
der Technischen Universität Darmstadt
zur Erlangung des Grades
eines Doktors der Naturwissenschaften
(Dr. rer. nat.)
genehmigte

Dissertation

von
Dipl.-Math. Niklas Benjamin Niemann

Referent	Prof. C. Herrmann
Koreferent	Prof. B. Kümmerer
Tag der Einreichung	14. Dezember 2006
Tag der mündlichen Prüfung	9. Februar 2007

Darmstadt 2007
D17

Bibliografische Information der Deutschen Nationalbibliothek

Die Deutsche Nationalbibliothek verzeichnet diese Publikation in der
Deutschen Nationalbibliografie; detaillierte bibliografische Daten sind
im Internet über http://dnb.d-nb.de abrufbar.

ISBN 978-3-8325-1563-8

Logos Verlag Berlin
Comeniushof, Gubener Str. 47,
10243 Berlin
Tel.: +49 030 42 85 10 90
Fax: +49 030 42 85 10 92
INTERNET: http://www.logos-verlag.de

Danksagung

Regular Rings, von Neumann and Murray,
they got it all started.
Operator algebras and projection lattices,
from there it all departed.
Irving Kaplansky,
various generalisations.
Berberian and Handelman,
embeddings and coordinatisations.
Representations of Rings,
Jacobson got it done.
His concept, primitivity, was the one.
For rings with involution,
there was quite some call.
Remember Herstein, Rowen, Wiegandt et al.
Regular rings and lattices,
different concepts of a frame.
Jónsson and others, that was their game.
In the new millennium, Micol and ∗-regularity,
results concerning coordinatisation and representability.
Christian Herrmann, insight, wisdom and perspectivity,
to bring it all together, in its whole variety.

3

Mein erster und besonderer Dank gilt Prof. Dr. Christian Herrmann. Ich danke ihm dafür, seit Beginn meines Studiums auf lebendige, inspirierende und mitreißende Weise *Mathematik* vermittelt zu haben. Ich danke ihm für seine außergewöhnliche Betreuung – für mich war und ist er der beste Betreuer der Welt.

Ich möchte ihm darüber hinaus dafür danken, in all den Jahren meines Studiums ein beständiger Begleiter meines mathematischen Lebensweges gewesen zu sein. Seine bemerkenswerten Veranstaltungen, bei denen man (manchmal erst, aber immer) noch Semester später merkte, was für wichtige Grundlagen gelegt und welche tiefgreifenden Zusammenhänge gelehrt worden waren, waren für mein Studium, für mich, von immenser Bedeutung. Ich danke ihm weiterhin für sein Entgegenkommen und Aufnehmen meiner Begeisterung für Darstellungstheorie nach meiner Rückkehr aus Irland; für seine Ausflüge in mathematische Gefilde ebenso wie für seine Geduld; dafür, dass er sich stets Zeit für seine Schützlinge genommen hat – und außerdem für das entgegengebrachtes Vertrauen in meine mathematischen Fähigkeiten.

Von ganzem Herzen möchte ich Daniela Engelbert danken. Ich danke ihr dafür, dass sie seit Beginn meiner Promotion bei mir, mit mir gewesen ist. Durch sie, ihre Gegenwart und ihre Liebe wurden die schönen Momente noch schöner, die schweren Tage etwas weniger schlimm. Darüber hinaus danke ich ihr dafür, dass sie mich und meine schwankenden Launen in dieser Zeit ertragen hat, dass sie immer an meiner Forschung interessiert gewesen ist, dass sie sich meine Probleme geduldig angehört und mit mir diskutiert hat, und dass sie im Endstadium meiner Arbeit diese, insbesondere meine eigenwilligen Formulierungen, mit grüner und blauer Tinte verziert hat.

Mein besonderer Dank gilt meinen Eltern – für ihre Liebe, für die Zeit mit ihnen, sowie für ihr Vertrauen, ihre Ermutigung und ihre Unterstützung während der gesamten Zeit meines Studiums.

Natürlich möchte ich allen weiteren Mitgliedern der Prüfungskommission danken:

Prof. Burkhard Kümmerer, dem Koreferenten meiner Arbeit. Ich danke ihm ausdrücklich für das Erstellen des zweiten Gutachtens. Seine Forschungstätigkeit im Bereich der Operatortheorie machte seine Wahl als Koreferenten beinahe kanonisch. Weiterhin danke ich ihm einerseits für seine Veranstaltungen zu Operator- und von Neumann-Algebren, in deren Rahmen ich einen Einblick in funktionalanalytische Betrachtungsweisen erhalten konnte, und andererseits für die Gelegenheit, in einer arbeitsgruppenübergreifenden Runde über die fachlichen und historischen Zusammenhänge unserer Arbeitsgebiete vorzutragen.

Prof. Martin Otto, dem arbeitsgruppeninternen Mitglied der Prüfungskommission. Als Professor der Arbeitsgruppe Algebra und Logik hatte er einen besonderen Zugang zu den in meiner Arbeit verwendeten Methoden der universellen Algebra und zu vielen der zugrundeliegenden Beweismechanismen. Ich danke ihm insbesondere für die fachliche Expertise: Prof. Otto suchte vor, während und nach der Prüfung die angeregte mathematische Diskussion und machte deutlich, wie manch trickreicher Beweis modelltheoretisch vereinfacht werden konnte.

Prof. Peter Spellucci, den ich mir insgeheim schon lange als Mitglied der Prüfungskommission erträumt hatte. Ich danke ihm dafür, dass er meine ungewöhnliche Bitte nicht abgeschlagen hat, ungeachtet des für ihn dadurch entstehenden Aufwandes. Außerdem bin ich ihm von Herzen dafür dankbar, dass er meine Prüfung durch seine Gegenwart, mit seiner (erwarteten) Frage nach Anwendungen, mit der daraus entstehenden Diskussion über (für das Publikum unerwartete) Themengebiete und mit seinem Humor ungemein bereichert hat.

Prof. Alexander Martin, dem Dekan und Leiter der Prüfungskommission, danke ich dafür, dass er trotz seines vollen Terminkalenders Zeit gefunden hat. Ihm möchte ich insbesondere meinen Dank dafür aussprechen, zu eben jener Zeit Dekan gewesen zu sein.

Mein ausdrücklicher Dank gilt außerdem Dr. Achim Blumensath. Ich danke ihm sowohl dafür, dass ich ihn in den letzten Jahren immer wieder mit Fragen zu Mathematik einerseits und zu LATEX und Layout andererseits belästigen durfte, als auch für das mühevolle Korrekturlesen meiner Arbeit, für seine zahllosen Anmerkungen und Hinweise. Außerdem danke ich ihm für die tatkräftige unterstützende Versorgung mit Literatur, sei es fachspezifische oder fachfremde.

Darüber hinaus möchte ich all jenen danken, die die Zeit meines Promotionsstudiums an diesem Fachbereich schöner, angenehmer, erträglicher, menschlicher gemacht haben. Leider ist es mir nicht möglich, all diese Menschen hier namentlich aufzuführen. Nennen möchte jedoch ins-

besondere die Folgenden: Die beiden Sekretärinnen der
AG 1/14, Frau Barbara Bergsträsser und Frau Ute Gal-
ter, die die Atmosphäre der Arbeitsgruppe durch ihre
Freundlichkeit bereichert haben, und die außerdem stets
bereit waren, mir diese oder jene Arbeit abzunehmen. Die
unglaublich hilfsbereiten und freundlichen Mitarbeiterin-
nen der Bibliothek des Fachbereiches, Frau Nicole Krüger
und Frau Kornelia Kernbach, die meine Vergesslichkeit
geduldig ertragen, mir immer wieder bei der Literatur-
recherche ausgeholfen und selbst meine etwas exotischen
Literaturwünsche gegen alle Widrigkeiten erfüllt haben.
Die Dekanatssekretärin Elke Roder, die besonders in der
Endphase meiner Promotion mit ihrer herzlichen Hilfs-
bereitschaft ein Lichtblick im grauen Beton war.

8

Prologue

I have a story to tell you.
It has many beginnings,
and perhaps one ending. Perhaps not.
Beginning and endings are
contingent things, anyway; inventions, devices.
Where does any story really begin?

— From THE ALGEBRAIST
by Iain M. Banks

The origins of von Neumann-regular rings and ∗-regular rings are the works of John von Neumann[1] and Francis J. Murray during the 30ies of the last century. They constitute a connection of the areas of operator theory, ring theory and lattice theory.

As pointed out by Goodearl [Good91], the most prominent connections between operator algebras and regular rings are the concepts of a regular ring associated with an operator algebra, a continuous geometry and the structural analogies between regular rings and operator algebras. Historically, the original purpose of von Neumann and Murray (see [MN]) was the wish to coordinatise pro-

---

[1] János Lajos Neumann

9

jection lattices of von Neumann-algebras. Here, the connection between operator theory and lattice theory is the following: On the one hand, a von Neumann-algebra A is a weakly closed self-adjoint algebra of bounded operators – on the other hand, its set of projections $P(A)$ constitutes a complete orthocomplemented lattice. For a finite von Neumann-algebra, von Neumann and Murray were able to construct a $*$-regular ring coordinatising $P(A)$.

The pioneer work of Murray and von Neumann was the basis for vast development. An approach focusing on the algebraic point of view was spurred by Kaplansky and his introduction of several generalisations of von Neumann-algebras [Kap68]. As mentioned in [Good91], AW^*-algebras and Baer-$*$-rings are of particular interest among these. While the former is defined to be a C^*-algebra with the additional requirement that the annihilator of any subset is generated by a projection, the latter is defined to be any involutive ring with just this additional postulation [Kap68]. For AW^*-algebras and Baer-$*$-rings, coordinatisation theorems similar to the one of Murray and von Neumann were proven by Berberian who extended the construction of von Neumann and Murray to finite AW^*-algebras and even to suitable Baer-$*$-rings [Ber57], [Ber72]. In the context of the larger class of Rickart-C^*-algebras, Handelman was able to show that any finite Rickart-C^*-algebra A embeds canonically into a regular ring coordinatising $P(A)$. For the interweavement of the theories of regular rings, operator algebras and lattices, see the introduction of the monograph of Goodearl [Good91], the survey article of Holland [Hol70] and of course, the works of von Neumann, Kaplansky, Berberian, Handelman et al. [MN],

[vN60], [Kap68], [Berb72], [Ber57], [Ber72].

Regular and ∗-regular rings were analysed by other well-known mathematicians: Ara and Menal [AM84] could show that every ∗-regular ring with unit is finite and thus were able to prove that every ∗-regular ring with unit satisfying the square root axiom is directly finite. Goodearl, Menal and Moncasi [GMM93] proved that regular rings are residually Artinian. The connection between ∗-regular rings and lattices was further developed by Halperin and Jónsson, working on questions of coordinatisation of complemented modular lattices.

Considering the historical origin again, one can speculate that von Neumann was inspired by both operator theory and lattice theory to introduce the notion of a ∗-regular ring: On the one hand, the requirement of positivity of the involution can be seen as the appropriate generalisation of the involution of operator algebras. On the other hand, ∗-regular rings give rise to a strong class of lattices which are closely connected to operator algebras.

This thesis shows that this speculation might have some substance, that is, the concept of a ∗-regular ring indeed gives an adequate axiomatic framework for regular rings of operators, if one is prepared to deal with vector spaces over general involutive skew fields, equipped with a scalar product. We show that every ∗-regular ring is representable in this sense, and that every variety of ∗-regular rings is generated by its simple Artinian members.

Furthermore, we consider the larger class of regular involutive rings and questions of their representability. In the context of rings without involution, Jacobson [Jac64] proved that representability as subrings of endomorphism rings of vector spaces is captured by primitivity. Dealing with involutive rings, one can introduce the notion of ∗-primitivity and representations in terms of bi-vector spaces, as done in [Row88] and [WieEtAl05]. Alternatively, one can examine primitive rings endowed with an involution, with the aim to construct an appropriate nondegenerated form on the vector space to capture the involution. Continuing the work done in [Mic03] and [Nie03], we give a complete characterisation of representability of regular involutive rings.

The thesis relies heavily on connections between regular rings and lattices, in particular, between ∗-regular rings and modular ortholattices, and lattice-theoretical results. In approach and technique, this thesis follows [Mic03], [Nie03], [HR99], [Herr], [HS].

In Chapter 1, we will introduce the necessary notation and terminology as well as tools from universal algebra.

Chapter 2 deals with representations in vector spaces of countable dimension and with row finiteness of matrix representations with respect to an appropriate basis. The section about questions of finiteness of representable rings is of particular interest.

Chapter 3 is concerned with the completion of the characterisation of representability of regular involutive rings, as started in [Nie03]. It will fill in the blank left by a result stated but not proven by Herstein and extend previous results to symplectic representations and representations over skew fields of characteristic two. Furthermore, the chapter studies the concept of approximation introduced by [Mic03] and atomic extensions of rings that admit a faithful linear representation.

Chapter 4 is devoted to *-regular rings. In this chapter, the main results of this thesis - every *-regular ring admits a faithful generalised positive representation and every variety of *-regular rings is generated by its simple Artinian members - are presented.

Chapter 5 deals with atomic regular involutive rings. We will show to what extend we can represent atomic regular involutive rings. In particular, we face the difficulty that subdirect irreducibility and *-subdirect irreducibility do not coincide in general regular involutive rings.

Another Prologue: The German Summary

Diese Arbeit ist der Untersuchung von regulären involutiven und *-regulären Ringen und der Frage nach ihrer Darstellbarkeit gewidmet. Sie kann als Fortführung von vorherigen von Prof. Dr. C. Herrmann betreuten Arbeiten, der Doktorarbeit von Florence Micol [Mic03] und meiner Diplomarbeit [Nie03], verstanden werden. Die zentralen Resultate dieser Arbeit sind die positive Klärung der Frage, ob jeder *-reguläre Ring in geeigneter Weise darstellbar ist, sowie der Beweis, dass jede Varietät von *-regulären Ringen von ihren einfachen artinschen Elementen erzeugt wird.

Wesentlich für die Resultate dieser Arbeit waren die starken Verbindungen zwischen *-regulären Ringen und Verbänden. Ansatz und Techniken sind ähnlich wie in [Mic03], [Nie03], [HR99], [Herr], [HS].

In Kapitel 1 werden Notation und Terminologie eingeführt, bekannte Beispiele und Grundlagen aus Ring- und Verbandstheorie genannt sowie Werkzeuge der universellen Algebra vorgestellt.

In Kapitel 2 werden Darstellungen von Vektorräumen von abzählbarer Dimension untersucht. Es wird gezeigt, dass in Bezug auf eine geeignete Basis die Matrixdarstellung eines stetigen Endomorphismus stets zeilenendlich ist. Außerdem werden Endlichkeitsfragen von darstellbaren Ringen untersucht.

Kapitel 3 vervollständigt die in [Nie03] begonnene Charakterisierung von Darstellbarkeit von regulären involutiven Ringen durch Erweiterung des dortigen Ansatzes auf symplektische Darstellungen und Darstellungen über Schiefkörpern von Charakteristik zwei. Die Problematik, dass ein Beweis in [Hst76] fehlerhaft war, wird durch das Heranziehen anderer Quellen behoben. Weiterhin werden der von Micol in [Mic03] vorgestellte Begriff der Approximierbarkeit sowie atomare Erweiterungen von linear darstellbaren Ringen untersucht.

Kapitel 4 ist $*$-reguläre Ringen und den Hauptergebnissen der Arbeit gewidmet: Jeder $*$-reguläre Ring besitzt eine treue positive G-Darstellung. Jede Varietät von $*$-regulären Ringen wird von ihren einfachen artinschen Elementen erzeugt.

Kapitel 5 beschäftigt sich mit atomaren regulären involutiven Ringen, insbesondere mit der Schwierigkeit, dass die Begriffe von subdirekt unzerlegbar und $*$-subdirekt unzerlegbar bei allgemeinen regulären involutiven Ringen nicht zusammenfallen. Es wird untersucht, inwieweit sich die Frage nach der Darstellbarkeit von atomaren regulären involutiven Ringen beantworten läßt.

Contents

1 Preliminaries

*The nice thing about standards is
that there are so many of them to choose from.*

— Andrew S. Tannenbaum

In this chapter, we will introduce the main objects of this thesis. At the beginning of the chapter, we establish notation and conventions used throughout the thesis. Subsequently, we sketch the ring-theoretical background necessary for this thesis and give a short introduction into the basics of regular rings. Afterwards, we deal with modular lattices and frame. The next sections are devoted to the connection between regular rings and modular lattices and our notion of a representation. Finally, we present the necessary tools of universal algebra.

The most important concepts in this chapter are *regularity of rings, frames in lattices* and *representations of rings and lattices*.

Motivation, historical references and examples will be provided in limited extend.

1.1 Notation and Conventions

The main objects considered in this thesis are rings and lattices. In the two subsequent sections, we introduce the notation used throughout the thesis.

Rings

In this thesis, the term *ring* will be used for rings with or without unit. If the considered ring contains a unit, we will state this explicitly (by using the term 1-(sub)ring). We denote rings by R, S, T, C.

For a ring without unit, we follow the notation of [Mic03] and use the term *plain ring*. In particular, when we consider a ring (with or without unit) and a subring explicitly not containing a unit, we will use the term *plain subring*. In the literature, one can encounter the terms *rng* and *subrng*, too (where the suggested pronunciation is *rŭng*) (see [Jac89], [Row88]).

Definition 1.1.1. Regular ring

A ring R will be called *(von Neumann-)regular* if for every element x in R there exists an element y in R such that $xyx = x$.

For $x \in R$, every element y satisfying $xyx = x$ will be called a *quasi-inverse of* x.

Example 1.1.1. Classical examples of regular rings are arbitrary products of skew fields. Other classical examples are matrix rings $M(n, D)$, where $n \in \mathbb{N}$ and D a skew field.

Definition 1.1.2. Involutive ring

An *involution* on a ring R is a unary operation $* : R \to R$ such that

1. $(a + b)^* = a^* + b^*$

2. $(a \cdot b)^* = b^* \cdot a^*$

3. $(a^*)^* = a$

i.e., an anti-automorphism of order 2. A ring equipped with an involution will be called an *involutive ring*.

If R is an involutive ring, the involution will be considered as part of the algebraic structure: While a ring is an algebraic structure of type

$$(R, +, \cdot, -, 0, 1) \quad \text{or} \quad (R, +, \cdot, -, 0)$$

an involutive ring is an algebraic structure of type

$$(R, +, \cdot, -, {}^*0, 1) \quad \text{or} \quad (R, +, \cdot, -, {}^*0).$$

Example 1.1.2. Classical examples for rings with involution are the field $(\mathbb{C}, {}^-)$ of complex numbers, with complex conjugation, the ring $(M(n, \mathbb{R}), {}^t)$ of all square matrices of fixed size n over the field of real numbers, with transposition, and the ring $(M(n, \mathbb{C}), {}^*)$ of all square matrices of fixed size n over the field of complex numbers, with adjunction.

A more general example is given by $(M(n, \mathbb{K}), {}^*)$, $\mathbb{K}$ an arbitrary (skew) field, with the involution given by transposition, possibly combined with an involution on $\mathbb{K}$, if $\mathbb{K}$ is an involutive (skew) field.

Remark 1.1.1. Note that, depending on the context, the *natural* involution on $M(n, \mathbb{K})$ might well be a different one than transposition. Notice furthermore that $M(n, \mathbb{K})$ gives rise to further examples, namely subrings closed under the involution.

Definition 1.1.3. Regular involutive ring
A regular ring R equipped with an involution $* : R \to R$ will be called a *regular involutive ring*.

Definition 1.1.4. ∗-Regular ring
Let R be a regular involutive ring. If in addition the involution $* : R \to R$ is positive, that is, if $xx^* = 0$ implies $x = 0$, then R will be called a *∗-regular ring*.

Example 1.1.3. The standard examples for regular involutive rings are full matrix rings over the real or the complex numbers with transposition or adjunction. Again, more general examples for regular involutive rings are full matrix rings over appropriate skew fields, equipped with transposition, possibly combined with an involution of the skew field. Of course, the standard examples $(M(n, \mathbb{R}),^t)$ and $(M(n, \mathbb{C}),^*)$ are both ∗-regular.

Definition 1.1.5. Idempotents and projections
In the context of involutive rings, we have to distinguish between idempotent elements and idempotent Hermitian elements. We use the term *idempotent* for an element x such that $x^2 = x$ and the term *projection* for an element x such that $x^2 = x^* = x$. We use the letters p, q for projections, while we use e, f, g for idempotents and projections.

In the literature, idempotents are sometimes called *skew projections*.

Lattices

In this thesis, the term *lattice* will be used for a partially ordered set with the binary operation of supremum (join) and infimum (meet). Throughout this thesis, we will denote the binary operations by $+$ and $\cdot$ (commonly used are also $\vee$ and $\wedge$), that is, considered as an algebraic structure, a lattice is of type $(L, +, \cdot)$.

All lattices considered in this thesis have a smallest element, denoted by 0. Most of them also have a greatest element, denoted by 1. By a *bounded lattice*, we mean a lattice with top and bottom. We use the notation $0(\text{-}1)$-lattice to indicate that the bottom (and the top) element are considered as part of the algebraic structure. In general, the bounds are not part of the signature and they need not be preserved by general lattice homomorphisms.

We use the notation $a \oplus b$ or $\bigoplus a_i$ for the sum of independent elements a and b or for the sum of a family $\{a_i : i \in I\}$ of independent elements.

An *interval* $[a, b]$ in a lattice L is a subset of L of the form $\{x : x \in L.\ a \leq x \leq b\}$, where $a, b \in L$. A *section* of a lattice is an interval of the form $[0, b]$, where $b \in L$. Intervals and sections are bounded lattices in their own right, with the lattice operations inherited from L.

We define the *height* $h(L)$ or the *dimension* $\dim(L)$ of a lattice L to be the supremum of all cardinalities $|C| - 1$, where C is a chain of L.

Definition 1.1.6. Modular lattice
A lattice L satisfying the modular law

$$x \leq y \Rightarrow y \cdot (x + z) = x + (y \cdot z) \qquad \text{(M)}$$

will be called a *modular lattice*.

Definition 1.1.7. Complemented lattice
A bounded lattice L such that to each element $a \in L$ there exists at least one $b \in L$ such that $a \oplus b = 1$ will be called a *complemented lattice*.

For $a \in L$, every element $b \in L$ satisfying $a \oplus b = 1$ will be called a *complement of a in L*.

In a *relatively* or *sectionally complemented* lattice, the existence of a complement is guaranteed in each interval $[a, b] \subseteq L$ or each section $[0, a] \subseteq L$. In particular, relatively or sectionally complemented lattices do not have to be bounded.

Definition 1.1.8. Modular ortholattice
A bounded modular lattice L equipped with an ortho-complementation $^{\perp} : L \to L$ will be called a *modular ortholattice*.

1.2 Prelude

As indicated at the beginning of the chapter, this thesis is not self-contained: It is neither self-contained nor detailed enough to be understandable without some background in ring theory (as well as lattice theory and universal algebra). In particular, the reader should be familiar with the following classes of rings: *Simple and semisimple rings, primitive and semiprimitive rings, prime and semiprime rings*, and *Artinian rings*.

This prelude is an attempt to introduce some of the basics that are used throughout the thesis and to present well-known concepts and results in a (hopefully) coherent way.

If the reader is willing to acquire more profound knowledge in ring theory, we recommend the following textbooks: **An Introduction to Ring Theory** [Cohn01], **Ring Theory** [Row88] and **A First Course in Noncommutative Rings** [Lam01]. Additionally, **Basic Algebra** (Volume I and II, [Jac85] and [Jac89]), **Structure of Rings** [Jac64] and **Rings of Operators** [Kap68] are suggested as concomitant reading.

For the reader to deal with regular rings, the book **Von Neumann Regular Rings** [Good91] and clearly, the various works of J. v. Neumann, are compulsory. I highly recommend the survey article **The current interest in orthomodular lattices** [Hol70] of S.S. Holland, Jr., since it illustrates the historical development and highlights the connections between the theory of operator algebras, regular rings and orthomodular lattices.

We begin with the following definitions.

Definition 1.2.1. Minimal ideal
We call a non-vanishing two-sided ideal of a ring *minimal* if it contains no other two-sided ideals of the ring except itself and the trivial ideal $\{0\}$. Likewise, we call a non-vanishing right (left) ideal of a ring *minimal* if it contains no other right (left) ideals of the ring except itself and the trivial ideal $\{0\}$.

Definition 1.2.2. Minimal element
Let R be a ring and e be an idempotent (a projection) in R. We call e *minimal* if the right ideal eR generated by e is minimal.

Now, we return to the structural notions mentioned above.

A ring R is called *simple* if R contains no non-trivial two-sided ideal. Following [vN60], we call a ring R *right semisimple* if R contains no non-trivial nilpotent right ideal, that is, if A is a right ideal of R with $A^n = \{0\}$, we can conclude that already $A = \{0\}$.

Remark 1.2.1. For simplicity, we agree to use the term *semisimple* instead of *right semisimple*.

Remark 1.2.2. As in [vN60], we note that if $A \neq \{0\}$ is a nilpotent right ideal, then A contains a nilpotent *principal* right ideal $B \neq \{0\}$. Hence, in defining right semisimplicity, we may restrict ourselves to *principal* right ideals.

A ring R is called *prime* if the product of two two-sided ideals vanishes only if one of the factors vanishes already. A ring R is called *semiprime* if the only nilpotent two-sided ideal in R is the trivial ideal $\{0\}$. An equivalent definition would be to call a ring semiprime if it is a subdirect product of prime rings (see [Row88], p. 164).

Lemma 1.2.3. *Let R be a prime ring and $r \in R$.*
Then Rr is a minimal left ideal iff rR is a minimal right ideal.

▷ Proof. [Row88, Proposition 2.1.27]. ◁

Lemma 1.2.4. *Let R be a semiprime ring.*
If A is a minimal left ideal A in R, then A has an idempotent generator e, i.e., $A = Re$ for some $e = e^2$. Furthermore, $B := eR$ is a minimal right ideal and eRe a skew field.

Conversely, if e is an idempotent such that eRe is a skew field, then Re is a minimal left and eR a minimal right ideal in R.

▷ Proof. See [Hst76, Lemma 1.2.1], or [Nie03], p. 61–63. ◁

A ring R will be called *(left or right) primitive* if there exists a faithful simple (left or right) module over R.

Remark 1.2.3. As stated in [Row88, Remark 2.1.14], every primitive ring is prime, whereas the converse does not hold, that is, not every prime ring is a primitive one. This implication does not even hold for commutative rings (in this context, see [Row88], p. 153–154). But the following holds.

Lemma 1.2.5. *Let R be a prime ring containing a minimal left (or a minimal right) ideal.*
Then R is primitive.

▷ Proof. [Row88, Proposition 2.1.15]. ◁

Remark 1.2.4. It might be considered as historical inconvenience (or poetic license), but the convention whether one considers left or right primitive rings varies from author to author, even from book to book. E.g., in [Hst76], Herstein considers right primitive rings, as does Jacobson in [Jac64] and accordingly, Micol in her thesis [Mic03]. On the other hand, in [Row88], Rowen focuses on left primitivity – and so does Jacobson in the second volume of his treatise **Basic Algebra** [Jac89].

We follow the notion of [Row88] and [Jac89], that is, we agree to use the term *primitive* for a ring R which is *left* primitive. As in [Row88], we use the notion of *simple* module instead of *irreducible* module. Consequently,

Jacobson's density theorem reads as follows (see [Row88, Theorem 2.1.6]):

Theorem 1.2.6. *(Jacobson's density theorem)*
Suppose that R has a faithful simple (left) module and let $D = End(_RM)$.[1]

Then R is dense in $End(M_D)$, M_D a right vector space over the skew field D.

Theorem 1.2.7. *(Wedderburn-Artin)*
Let R be a primitive ring satisfying the descending chain condition on left ideals.

Then R is isomorphic to $M_n(D)$ for a suitable skew field D and a suitable $n \in \mathbb{N}$.

▷ Proof. See [Row88, Theorem 2.1.8]. ◁

Lemma 1.2.8. *Let R be a primitive involutive ring and p a projection decomposing into a sum $p = a + b$ of two commuting minimal non-Hermitian idempotents a, b with $ab = 0$.*
    Then $S := pRp \cong M(2, D)$, for some skew field D.

▷ Proof. It is easy to show that S is primitive (e.g., take the criterion given in [Row88], Proposition 2.1.11). Hence, $S \cong M(n, D)$ for some skew field D, where a suitable $n \in \mathbb{N}$ is $n = 2$. ◁

Following [Row88], p. 179 and [Jac89], we call a ring R *semiprimitive* if R is a subdirect product of primitive rings, i.e. if the intersection of all primitive ideals of R vanishes, i.e. if the Jacobson radical of R is trivial. In

---

[1]By Schur's Lemma, D is a skew field.

the literature (e.g., in [Lam01]), semiprimitive rings are sometimes called *J(acobson)-semisimple.*

Following [Row88], p. 22 and p. 167, we call a ring R *left Artinian* (resp. *right Artinian*) if R is Artinian as left (resp. right) module over itself, i.e. if R satisfies the descending chain condition on left (right) ideals, i.e. if R satisfies the minimum condition on left (right) ideals. A ring is called *Artinian* if it is both left and right Artinian.

As in [Row88], Section 2.3, we have the following.

Theorem 1.2.9. *The following conditions are equivalent for a ring R with unit.*

1. *R is a finite direct product of simple Artinian rings.*

2. *$R = Soc(R)$.*

3. *R is semiprime and left Artinian.*

Here, the socle $Soc(R)$ of R is defined to be the sum of all minimal left ideals of R. (See [Row88], p. 156. In particular, $Soc(R)$ is taken to be $\{0\}$ if there are none.)

▷ Proof. [Row88, Theorem 2.3.10]. ◁

Furthermore, the following holds.

Lemma 1.2.10. *Every semiprime Artinian ring has a unit.*

▷ Proof. [Row88, Exercise 2.3.7]. ◁

1.3 Regular Rings

In Section 1.1, we have introduced the notion of a *regular ring*. As pointed out in the Preface, the concept of *regularity* was invented in the mid 1930ies by John von Neumann and F.J. Murray. Their motivation for this invention had several sources. John von Neumann himself reasoned in his article [vN36] that regular rings are a natural generalisation of division algebras: The requirement that in a regular ring, every element element has a quasi-inverse is remarkably similar to the requirement that in skew field, every element has an inverse: We have

$$\forall x.\exists y.\ xyx = x \quad \text{vs.} \quad \forall x.\exists y. \quad xy = yx = 1.$$

Furthermore, John von Neumann wanted to generalise the notion of semisimplicity. Of course, another motivation was the close connection between regular rings and modular lattices. Before we elucidate the two latter motivations in 1.3 and 1.5, we introduce some basic results concerning regular rings.

The following characterisation of regular rings was proven by von Neumann and can be found in [vN36] and [Good91].

Theorem 1.3.1. *For a ring R, the following conditions are equivalent:*

1. *R is regular.*

2. *Every principal right (left) ideal of R is generated by an idempotent.*

3. *Every finitely generated right (left) ideal of R is generated by an idempotent.*

Remark 1.3.1. Construction principles As pointed out by Goodearl [Good91], the class of regular rings is obviously closed under homomorphic images, direct products and direct limits.

Brown and McCoy introduced the following notion of a regular two-sided ideal.

Definition 1.3.2. Regular ideal
A two-sided ideal I in a ring R is *regular* if for every element $x \in I$ there exists an element $y \in I$ such that $xyx = x$, that is, if I is regular as a ring.

As in [Mic03, Lemma 2.13], we have the following.

Lemma 1.3.3. *Let I and K be two-sided ideals in a ring R such that $I \leq K$.*

Then K is regular iff I and K/I are both regular. In particular, every two-sided ideal in a regular ring is regular.

▷ Proof. [Good91, Lemma 1.3]. ◁

Lemma 1.3.3 can be used to show the following.

Proposition 1.3.4. *Any finite subdirect product of regular rings is regular.*

▷ Proof. [Good91, Proposition 1.4]. ◁

Remark 1.3.2. Note that a subdirect product of infinitely many regular rings need not be regular. But there are other construction principles.

Proposition 1.3.5. *Let R be a regular ring.*

Then the following rings are regular:

1. *eRe for every idempotent e in R.*

2. *M(n, R) for every natural number n.*

3. *$End_R(A)$ for every finitely generated projective module over R.*

▷ **Proof.** See [Row88] and [Good91, Theorem 1.7]. The last statement is due to John von Neumann. ◁

In [Good91], Goodearl points out the limits of such general principles: In contrast to Proposition 1.3.5 (3), not every ring which is associated with regular rings in some *reasonable* manner is necessarily regular. In particular, endomorphism rings of more general modules over regular rings need not be regular. Goodearl even gives an example of a finitely generated module over a *commutative* regular ring R such that $End(A)$ is not regular [Good91, Example 1.9].

For commutative rings, we have the following results.

Theorem 1.3.6. *For a commutative ring, the following are equivalent:*

1. *R is regular.*

2. *R_M is a field for all maximal ideals M of R.*

3. *R has no nonzero nilpotent elements, and all prime ideals of R are maximal.*

4. *All simple R-modules are injective.*

Lemma 1.3.7. *Every commutative subdirectly irreducible regular ring is a field.*

▷ Proof. Consider $x \neq 0$ in R. Let y be a quasi-inverse of x. Then xy is a central idempotent in R, giving rise to a direct decomposition of R. Since R is subdirectly irreducible, we can conclude that either $1 - xy$ or xy vanishes. Since $x \neq 0$, we have $xyx = x \neq 0$. Consequently, $1 - xy = 0$. ◁

This result can also be found in [Lam01, Chapter 7].

Regular Rings and the Prelude

In this section, we want to relate the ring-theoretical properties that we introduced in Section 1.2 to the concept of von Neumann-regularity. As a start, we address the before-mentioned motivation of John von Neumann to view *regularity* as a generalisation of *semisimplicity*. In [vN60] and [vN36], he points out the connections between regularity, semisimplicity and chain conditions. First, he stresses that his notion of *semisimplicity* coincides with the definition of *rings without radical* given in [vdWII]. In contrast to this convention of von Neumann, other authors use the term *semisimple* for rings without radical that satisfy the descending chain condition also. For example, see [vdWII] or [Row88]. In the latter one, Rowen calls a ring satisfying any of the equivalent conditions of Theorem 1.2.9 *semisimple Artinian* or just *semisimple*.

John von Neumann (see [vN60]) then emphasis the following two results.

Proposition 1.3.8. *Every regular ring is semisimple in the sense that it contains no non-trivial (principal) right ideal.*

Proposition 1.3.9. *Every semisimple ring satisfying the descending chain condition is regular.*

As an immediate consequence of regularity, we get semiprimeness.

Lemma 1.3.10. *Each regular ring is semiprime.*

▷ Proof.　Let I be a two-sided ideal in R and $x \in I$, $x \neq 0$. Since R is regular, there exists a quasi-inverse y of x in R. Since I is an ideal, we have $yx \in I$, hence $x = xyx = x(yx) \in I^2$. Consequently, $I^2 \neq \{0\}$. ◁

Hence, we get the following.

Corollary 1.3.11. *Every subdirectly irreducible regular ring is prime.*

Lemma 1.3.10 and Corollary 1.3.11 can be combined with Lemma 1.2.5 to the following consequence.

Corollary 1.3.12. *Let R be a subdirectly irreducible regular ring containing a minimal left (or a minimal right) ideal.*
Then R is primitive.

But in fact we even have the following result.

Proposition 1.3.13. *A von Neumann-regular ring is semiprimitive.*

▷ Proof. [Jac89, Exercise 4.2.5]. ◁

Hence, we have arrived at the following result, given in [Lam01, Corollary 4.24].

Corollary 1.3.14. *For a ring, the following implications of properties hold.*

$$Artinian\ Semisimple\ \Rightarrow\ Regular\ \Rightarrow\ Semiprimitive$$

In particular, we can conclude that a subdirectly irreducible regular ring is primitive.

Remark 1.3.3. We close this introduction with a remark to minimal one-sided ideals. Of course, a minimal one-sided ideal is generated by a single element. As stated in Lemma 1.2.4 and [Nie03, Lemma 4.3.6], a minimal one-sided ideal in a semiprime ring is generated by an idempotent element. In particular, this is the case for a primitive ring.

Let R be a semiprime, a primitive, or a regular ring and I a one-sided ideal in R. The property of I to be minimal can be expressed in terms of the elements of the rings only. Consequently, the property of a semiprime, a primitive, or a regular ring to have a minimal one-sided ideal can be expressed in first-order logic.

∗-Regular Rings and the Prelude

The previous results can be combined to get the following.

Lemma 1.3.15. *Every subdirectly irreducible Artinian ∗-regular ring is simple.*

1.4 Modular Lattices

In this thesis, we deal mainly with *modular* lattices. Of particular interest are *complemented modular, relatively or sectionally complemented modular lattices* and *modular ortholattices*. We use the abbreviations CML, Rel-CML, SectCML, MOL.

We begin this section with the notion of *perspectivity* between elements of a lattice and with the introduction of several different notions of a *frame* in a lattice. Both concepts are important in the context of modular lattice theory and geometry. They date back to the 1930ies. For background literature, please consult the works of J. von Neumann [vN60], the survey of A. Day given at the Proceedings of the Fourth International Conference held at Puebla, Mexico in 1982 [Day82] and the works of R. Freese, B. Jónsson, C. Herrmann and A. Huhn.

Definition 1.4.1. Perspectivity
Let L be a modular lattice. Two elements a and b of L are called *perspective* (to each other) if there exists an element c such that $a \oplus c = b \oplus c$. If so, there exists such an element c with $c \leq a + b$. If a and b are perspective, we write $a \sim b$.

An element c that establishes the perspectivity between a and b is called an *axis of perspectivity* between a and b. Clearly, an axis of perspectivity between a and b is an axis of perspectivity between b and a, too.

Definition 1.4.2. Subperspectivity
Let L be a modular lattice and $a, b \in L$. We say that a is *subperspective to b* or *perspective to a part of b* if there exists $d \leq b$ such that a is perspective to d. If a is

subperspective to b, we write $a \lesssim b$.

An element c that establishes the subperspectivity between a and b is called an *axis of subperspectivity* between a and b. The part $d \leq b$ such that $a \sim d$ is called the *image of a under the subperspectivity between a and b*.

Lattices and Frames

There exist different notions of a *frame*. In [vN60], John von Neumann defined a *homogeneous basis* for a CML L (p. 93) and a *(normalised) system of axes of perspectivity* for a given homogeneous basis (p. 118). The combined system was called a *(normalised) frame* for L. Equivalently, G. Bergmann and A. Huhn introduced the notion of a *n-frame* (originally, a $(n-1)$-*diamond*) in a modular lattice (see the survey articles [Day82], [Day84] or the article of C. Herrmann in memory of A. Day [Herr95]).

The notion of a frame was subject to further development and generalisation. See [Jón60] for the introduction of a *partial frame*, a *large partial frame* and a *global frame*. In [Jón60], Jónsson defined a *large partial n-frame* in a bounded modular lattice B to be a subset of B consisting of independent elements $a_0, \ldots, a_{n-1}, d$ and the entries of a symmetric matrix $c = (c_{ij})_{i,j<n}$ such that the supremum of $a_0, \ldots, a_{n-1}$ and d equals the unit element 1_B, d consists of a sum of finitely many elements each of which is subperspective to a_0, and c_{ij} is an axis of perspectivity between a_j and a_i.

We adapt the definition of Jónsson in the following way: Decomposing d into k summands, each of which is subperspective to a_0, we incorporate these summands and their axes of subperspectivity to a_0 into the frame. Furthermore, we demand that the spanning elements of the frame are independent. Formally, we have

Definition 1.4.3. Large partial (n, k)-frame
A *large partial frame of format* (n, k) in a bounded modular lattice L is a subset

$$\Phi := \{a_i, a_{0i} : 0 \leq i < n + k\} \subseteq L$$

such that the following conditions are satisfied.

1. $\displaystyle\bigoplus_{i < n+k} a_i = 1_L.$

2. $a_0 + a_i = a_0 \oplus a_{0i} = a_i \oplus a_{0i}$ for $i = 1, \ldots, n - 1$.

3. $a_0(a_i + a_{0i}) + a_i = a_i \oplus a_{0i} = a_0(a_i + a_{0i}) \oplus a_{0i}$ for $i = n, \ldots, n + k - 1$.

That is, Φ contains $n + k$ independent elements a_i spanning the lattice L (condition (1)). Conditions (2) and (3) state that $a_1, \ldots, a_{n-1}$ are perspective to a_0 and $a_n, \ldots, a_{n+k-1}$ are subperspective to a_0, where the axes of (sub)perspectivity are just the a_{0i}. In particular, we have $a_0 \cdot a_{0i} = a_i \cdot a_{0i} = 0$ for all i.

The axes of perspectivity between a_i, a_j for indices $i, j < n$ can be constructed via the axes a_{0i}, a_{0j}: We have

$$a_{ji} = (a_{0j} + a_{0i}) \cdot (a_j + a_i)$$

and consequently

$$a_{ki} = (a_{kj} + a_{ji}) \cdot (a_k + a_i) \text{ for } i, j, k < n.$$

Likewise, we can construct the axis of subperspectivity a_{ji} between a_i and a_j for indices i, j such that $j < n$ and $n \leq i < n + k$.

For short, we call Φ a *large partial (n, k)-frame* or a *(n, k)-frame*, dropping the attribute *large partial* for the ease of notation and to avoid confusion with the notion of a large partial n-frame in the sense of Jónsson.

The basic example of a frame is the following.

Example 1.4.1. Let R be a ring with unit and n a natural number. Let $L(R_R^n)$ be the submodule lattice of the right module R_R^n. Denote the standard basis of R^n by $\{e_i : 0 \leq i < n\}$.

A *global n-frame* (see [Jón60]) in $L(R_R^n)$ is given by

$$\{a_i := e_i R : 0 \leq i < n\}$$

together with the axes of perspectivity

$$c_{ij} := (e_i - e_j)R = (e_j - e_i)R.$$

Lemma 1.4.4. *Let L be a CML and $a_0, a, b \in L$ such that $a_0 \leq a$, $a_0 \cdot b = 0$, and b is subperspective to a_0.*

Then the relative complement d of ab in $[0, b]$ is subperspective to a_0 and $a \oplus d = a + b$.

▷ **Proof.** Restrict the axis of subperspectivity between b and a_0 to d to get the subperspectivity between d and a_0.

Furthermore,

$$d + a = d + (ab + a) = (d + ab) + a = b + a$$

and
$$d \cdot a = (d \cdot b) \cdot a = d \cdot (a \cdot b) = 0$$

◁

Lemma 1.4.5. *Let L be a CML. If L contains a large partial n-frame in the sense of Jónsson, then L contains a (n, k)-frame.*

▷ **Proof.** All we have to show is that we can choose the summands of d in the definition of Jónsson such that all a_i, $0 \leq i < n + k$ are independent. More explicitly, we have to construct elements $a_i, n \leq i < n + k$ such that each one of them is subperspective to a_0, the elements $a_0, \ldots, a_{n+k-1}$ are independent, and $\sum a_i = 1_L$.

We do so by an inductive process. Let b be an element of L being subperspective to a_0 and assume that $b \cdot a \neq 0$, where $a := \sum_{k<m} a_k$ is the supremum of the already constructed independent part of the frame[2]. That is, we consider the following situation.

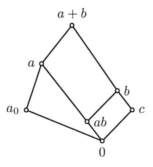

---

[2]Note that at the beginning of the inductive process, we have $m = n$ and $a = a_0 + a_1 + \cdots + a_{n-1}$

where c is a complement of $a \cdot b$ in the interval $[0, b]$. We apply Lemma 1.4.4: We set $a_m := c$ and take a_{0m} to be the restriction of the subperspectivity between b and a_0 to $c \leq b$. Then we have that $a + a_m = a + b$ and a_m is subperspective to a_0 via a_{0m}. ◁

Next, we introduce the concept of a stable frame. The main difference is that we incorporate all the axes of (sub)perspectivity (see Definition 1.4.3) and a set of relative complements.

Definition 1.4.6. Stable (n, k)-frame
Let L be a CML. A subset

$$\begin{aligned} \Phi \;=\; & \{a_i, a_{ij} : 0 \leq i < n, 0 \leq j < n + k\} \\ & \cup \{z_{ij} : j < n, n \leq i < n + k\} \subseteq L \end{aligned}$$

will be called a *stable (n, k)-frame* Φ *in* L if

1. $\{a_i, a_{0i} : 0 \leq i < n + k\}$ is an (n, k)-frame in L

2. for $i, j \in I$, $i < n$, a_{ij} is the axis of (sub)perspectivity between a_j and a_i

3. for each pair (i, j) of indices with $j < n$ and $n \leq i < n + k$, the element z_{ij} is a complement of b_{ji} in $[0, a_j]$, where b_{ji} is the image of a_i under the subperspectivity a_{ji} between a_i and a_j.

Lemma 1.4.7. *Let L be a CML.*
If L contains an (n, k)-frame, then L contains a stable (n, k)-frame.

▷ **Proof.** Choose the necessary relative complements. ◁

Definition 1.4.8. Orthogonal (n, k)-frame
Let L be a MOL. An (n, k)-frame Φ in L is called an *orthogonal (n, k)-frame* if the following additional condition is satisfied:

$$\forall i \in \{0, \ldots, n + k - 1\}. \qquad a_i^\perp = \sum_{j \neq i} a_j$$

Definition 1.4.9. Stable orthogonal (n, k)-frame
Let L be a MOL. A *stable orthogonal (n, k)-frame* is a stable frame such that

1. Φ as a frame satisfies the condition of Definition 1.4.8, and

2. the relative complements z_{ij} are relative orthocomplements.

Orthogonalisation via Jónsson

In this section, we want to show that for modular ortholattices, the notion of a (stable) orthogonal (n, k)-frame is the appropriate concept of a frame, that is, we want to show that we can orthogonalise a given frame. We will base the proof on arguments and results presented in [Jón60]. In fact, one could choose an alternative approach via ideas of Fred Wehrung, presented in [Weh98], using the notion of a *normal equivalence* in a modular lattice and the concept of a *normal* modular lattice.

Lemma 1.4.10. *Let L be a MOL and a, b projective elements in L.*
Then there exist four elements b_0, b_1, b_2, b_3 in L such that b is the direct orthogonal sum of b_0, b_1, b_2, b_3 and

each b_i is perspective to a part of a.

▷ **Proof.** The line of arguments follows Jónsson's proof of Lemma 1.4 in [Jón60]. The only real difference[3] is that we choose the relative complements in Jónsson's proof to be relative orthocomplements in the considered intervals. ◁

Lemma 1.4.11. *Let L be a MOL and a_0, a, b elements in L such that $a_0 \leq a$, $a \cdot b = 0$, and $b \lesssim a_0$.*

Then b decomposes into a direct orthogonal sum of five elements b_0, b_1, b_2, b_3, b_4 such that each b_i is subperspective to a_0.

▷ **Proof.** We choose $c := b \cdot a^\perp$ and d as the relative orthocomplement of c in $[0, (a+b) \cdot a^\perp]$. Clearly, as part of b, c is subperspective to a_0.

We claim that d is projective to a part $x \leq a_0$. To prove this claim, we set

$$r := a + c \quad \text{and} \quad w := r \cdot b$$

and choose a relative complement u of w in $[0, b]$. We calculate the following.

$$d + r = d + a + c = a + (d + c) = a + (a+b)a^\perp = a + b$$
$$u + r = u + a + c = a + b$$
$$u \cdot r = (u \cdot b) \cdot r = u \cdot w = 0$$

Thus, we have

$$u + r = d + r \quad \text{and} \quad u \cdot r = d \cdot r = 0.$$

---

[3] Another difference is that we have interchanged the roles of a and b.

Hence, d is perspective to $u \leq b$. Since b is subperspective to a_0, we can conclude that d is projective to a part x of a_0.

Using Lemma 1.4.10, we can decompose d into the direct orthogonal sum of 4 elements $b_1, \ldots, b_4$, each of which is subperspective to a_0. Together with $b_0 := c$, we have the desired result. ◁

Lemma 1.4.12. *Let L be a simple MOL and $a, b \in L$ non-trivial independent elements.*

Then there exist non-trivial elements a_0, b_0 with $a_0 \leq a$ and $b_0 \leq b$ such that a_0 and b_0 are perspective to each other. In particular, this holds if $b \leq a^{\perp}$.

▷ **Proof.** Since L is simple, the neutral ideal generated by a is the whole lattice. Then b is the sum of finitely many elements, each of which is perspective to a part of a. Choose one such non-trivial summand as b_0 and the corresponding perspective part of a as a_0. ◁

Lemma 1.4.13. *Let L be a simple MOL with $h(L) \geq n$.*

Then there exists a large partial n-frame Φ (in the sense of Jónsson) in L such that the first n elements $a_0, \ldots, a_{n-1}$ of Φ are orthogonal, that is, we have

$$a_k \leq \left(\bigoplus_{i < k} a_i \right)^{\perp}.$$

▷ **Proof.** Start with an element $a \neq 0, a \neq 1$. Since L is simple, there exist perspective elements b_0, b_1 with $b_0 \leq a$ and $b_1 \leq a^{\perp}$.

Assuming that $h(L) > 2$, we proceed by an inductive process. Let $k < n$ and assume that we have already constructed k independent pairwise perspective elements $b_0, \ldots, b_{k-1}$ such that $b_l \leq (\oplus_{i<l} a_i)^\perp$ for all $l \leq k-1$.

We have to distinguish between the two following cases.

$$\sum_{i<k} b_i = 1 \quad \text{and} \quad \sum_{i<k} b_i \neq 1$$

In the first case, since $h(L) \geq n$ and $n > k$, we can find elements $\tilde{b}_i$ with $\tilde{b}_i \leq b_i$ such that the $\tilde{b}_i$ are pairwise perspective and $\sum \tilde{b}_i < 1$. Therefore, we can focus on the second case. Consider $a := \oplus_{i<k} b_i$ and $a^\perp$. Since L is simple, there exist elements $c_0 \leq b_0$ and $c_1 \leq a^\perp$ such that c_0 and c_1 are perspective to each other. Using the perspectivities between the b_i, we can construct elements $a_0, \ldots, a_k$ with

$$a_0 := c_0 \leq b_0 \quad a_1 \leq b_1 \quad \ldots \quad a_k := c_1 \leq a^\perp$$

such that the a_i are independent and pairwise perspective and for all $l \leq k$, we have $a_l \leq (\oplus_{i<l} a_i)^\perp$. Continuing this procedure, we obtain n independent pairwise perspective orthogonal elements $a_0, \ldots a_{n-1}$. Since L is simple, each element in L is a sum of finitely many elements each of which is perspective to a part of a_0 [Jón60, Corollary 1.6]. In particular, for $1 \in L$, we get

$$1 = \bigoplus_{i<n} a_i \oplus d \quad \text{and} \quad d = \sum_{j<k} \alpha_j,$$

where each α_j is subperspective to a_0. ◁

Lemma 1.4.14. *Let L be a MOL containing an (n, k)-frame Φ such that $a_0, \ldots a_{n-1}$ are orthogonal, that is, if* *for all $k < n$, we have $a_k \leq (\oplus_{i<k} a_i)^{\perp}$.*

Then L contains an orthogonal (n, k')-frame for some k'.

▷ **Proof.** For $j \in \{n, \ldots, n + k - 1\}$, we consider the elements a_j inductively. By Lemma 1.4.10, we can decompose each a_j into at most four elements, each of which is subperspective to a_0, such that the four new summands contribute to an orthogonal sum. ◁

Corollary 1.4.15. *Let L be a simple MOL of height at least n.*

Then L contains a stable orthogonal (n, k)-frame, for some $k \in \mathbb{N}$.

Projectivity of Frames

In this section, we want to show that stable orthogonal frames of format (n, k) are projective. First, we consider large partial frames of format (n, k).

Lemma 1.4.16. *Let K and L be CMLs, $f : K \twoheadrightarrow L$ a surjective 0-1-lattice homomorphism and $\Phi \subseteq L$ a large partial (n, k)-frame in L.*

Then there exists a section $M \leq K$ and a set $\Psi \subseteq M$ such that

1. *$f_{|M} : M \to L$ is a surjective lattice homomorphism,*

2. *Ψ is a large partial frame in M of the same format as Φ, and*

3. *$f[\Psi] = \Phi$.*

Remark 1.4.1. It is folklore that global frames (or von Neumann-frames) are projective. In the following, we will sketch a possible inductive construction for large partial (n, k)-frames.

▷ Proof. The beginning of the inductive process is trivial. So, let $a_0, \ldots, a_{n-1}$ be the first n independent elements of the large partial (n, k)-frame Φ in L (i.e., $a_0, \ldots, a_{n-1}$ belong to the *frame* part of Φ). Let m be an index such that $m - 1 \leq n - 2$. Assume that for the first m elements $a_0, \ldots, a_{m-1}$ and the corresponding axes of perspectivity a_{im}, we have chosen elements $b_0, \ldots, b_{m-1}$ and b_{im} in K such that the following conditions are satisfied.

1. $f(b_i) = a_i$.

2. $f(b_{im}) = a_{im}$.

3. $\prod_{i=0}^{m-1} b_i = c$ and $f(c) = 0$.

4. $\sum_{i=0}^{m-1} b_i = d$ and $f(d) = \bigoplus_{i=0}^{m-1} a_i$.

Since K is a complemented modular lattice, we can assume that $c = 0$: If not, we replace the chosen elements with their intersection with the relative complement of c in $[0, d]$. Hence, the elements b_i, b_{ij} with $0 \leq i, j \leq m - 1$ form an m-frame in the section $[0, d]$. Nevertheless, we will speak of an element c.

Now consider the $(m + 1)$th element a_m. We choose a preimage $b_m \in K$. Without loss of generality, we can assume that $b_m \cdot d = 0$: Else, we replace b_m by a relative complement of $b_m \cdot d$ in the interval $[0, b_m]$.

Now we chose a preimage A of a_{0m}. Then the situation in L and K is the following.

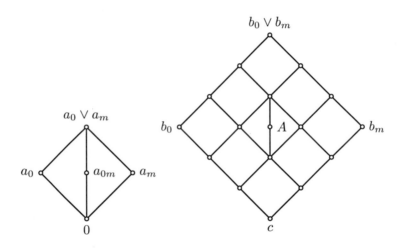

We agree on the following settings.

$$M := b_0 \cdot A \qquad N := b_m \cdot A$$

$$R := (A + b_m) \cdot b_0 \qquad S := (A + b_0) \cdot b_m$$

Furthermore, we choose a relative complement U of M in $[c, R]$ and a relative complement V of N in $[c, S]$ to get the following picture in K.

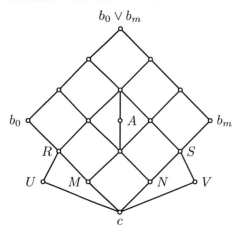

In this situation, we calculate

$$f(R) = (f(A) + f(b_m)) \cdot f(b_0) = (a_{0k} + a_k) \cdot a_0 = a_0$$

and similarly, $f(S) = a_k$. Furthermore, we note that $f(U)$ is a relative complement of $f(M)$ in $[f(c), f(R)]$. But since $f(c) = 0$, $f(R) = a_0$, and

$$f(M) = f(b_0) \cdot f(A) = a_0 \cdot a_{0m} = 0,$$

we have that $f(U) = a_0$. Likewise, we get $f(V) = a_m$.

Hence, applying the morphism f to $Q := (U + V) \cdot A$, we get

$$f(Q) = (a_0 + a_m) \cdot a_{0m} = a_{0m},$$

that is, Q is a preimage of a_{0m}.

We set

$$\tilde{c} := M + N = Ab_0 + Ab_m$$

and choose $\tilde{b}_0 := U$ and $\tilde{b}_m := V$ as the new preimages of a_0 and a_m.

We claim that Q is an axis of perspectivity between $\tilde{b}_0$ and $\tilde{b}_k$ in $[\tilde{c}, \tilde{b}_0 \vee \tilde{b}_k]$.

▷ **Proof.** Since $Q = (U + V)A$, we have $UQ \leq UA = 0$, because U was chosen as a relative complement of M in $[c, R]$. Likewise, $VQ = 0$.
On the other hand, we have

$$U + Q = U + (U + V)A = (U + V)(U + A) = (U + V),$$

because $U + A = U + M + A = R + A \geq V$ and so $U + A \geq U + V$. ◁

Now, we have to adjust this change in the elements $b_1, \ldots, b_{k-1}$ as well. To do so, we exchange b_j by

$$\tilde{b}_j := (b_{0j} + \tilde{b}_0) \cdot b_j.$$

A straightforward calculation shows that

$$f(\tilde{b}_j) = (a_{0j} + a_0) \cdot a_j = a_j,$$

that is, $\tilde{b}_j$ is a preimage of a_j, as desired.
Finally, we choose new axes $\tilde{b}_{ij}$ accordingly. By Remark 1.4.1, we can drag the chosen elements into a section $[\tilde{c}, d + b_m]$ of K. Completing the inductive step, we drop the tildes.

Having completed the inductive process for the *frame* part of Φ, we handle the *partial* part, i.e. the summands α_i of a^*, in a similar way. We consider the picture

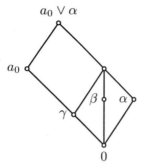

where $\alpha \in L$ is a summand of a^*, β the axis of subperspectivity and γ the corresponding part of a_0, i.e., α is subperspective to γ via β and $\gamma \leq a_0$. As in the previous procedure, we can assume that $\alpha \wedge a_0 = 0$.

We have to construct preimages r, s, t of α, β, γ such that r is subperspective to t via s and $t \leq b_0$. Focusing on the elements α, β, γ, we are in a similar situation as above, when we considered the *frame* part of ϕ. In the same way as above, we can choose preimages r, s, t such that $f(r) = \alpha, f(s) = \beta, f(t) = \gamma$, and t is the necessary axis of perspectivity between r and s. Again, we handle necessary adjustments as before. ◁

Similarly, we have the following.

Lemma 1.4.17. *Let K, L be CMLs, $f : K \twoheadrightarrow L$ a surjective 0-1-lattice homomorphism and $\Phi \subseteq L$ a stable (n, k)-frame in L.*

Then there exists a section $M \leq K$ and a set $\Psi \subseteq M$ such that the following hold.

1. *$f_{|M} : M \to L$ is a surjective lattice homomorphism.*

2. *Ψ is a stable frame in M of the same format as Φ.*

3. $f[\Psi] = \Phi$.

▷ **Proof.** The only thing left to show is that we can incorporate the choice of the necessary relative complements in the inductive procedure. For this, it is enough to show the following.

Claim 1.4.1. Let a and b be elements in K such that $b \le a$ and $f(b) \oplus c = f(a)$ for some $c \in L$. Then there exists $d \in K$ such that $b \oplus d = a$ and $f(d) = c$.

▷ **Proof.** Choose a preimage c_0 of c in K. Without loss of generality, we can assume that $c_0 \in [0, a]$. Then we intersect c_0 with b and choose a relative complement c_1 of bc_0 in $[0, c_0]$. Finally, we choose d as a relative complement of $b + c_1$ in $[c_1, a]$. That is, we are in the following situation in K.

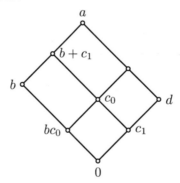

Since $f(c_1)$ is a relative complement of $f(bc_0)$ in $[0, c]$, we have $f(c_1) = c$. Likewise, we get $f(d) = c$. Hence, d is the desired relative complement of b in $[0, a]$. ◁ ◁

Lemma 1.4.18. *Let K, L be MOLs, $f : K \twoheadrightarrow L$ a surjective 0-1-lattice homomorphism and $\Phi \subseteq L$ a stable orthogonal (n, k)-frame in L.*

Then there exists a section $M \leq K$ and a set $\Psi \subseteq M$ such that

1. *$f_{|M} : M \to L$ is a surjective lattice homomorphism,*

2. *Ψ is a stable orthogonal frame in M of the same format as Φ, and*

3. *$f[\Psi] = \Phi$.*

▷ **Proof.** It is left to show that we can incorporate the orthogonality into the inductive process.

Therefore, assume that we have already chosen preimages $b_0, \ldots, b_{m-1}$ and b_{ji} of $a_0, \ldots, a_{m-1}$ and a_{ji} such that the elements $b_0, \ldots, b_{m-1}$ and b_{ji} constitute a stable orthogonal (r, s)-frame Ψ in the section $[0, d]$, where $d = \sum_{j=0}^{m-1} b_j$ and the format (r, s) with $r + s = m$ of Ψ is less than (n, k).

We choose a preimage b_m of a_m. In this situation, we can assume that $b_m \leq d^\perp$: Else, we replace it by $b_m \cdot d^\perp$. This choice already enlarges our orthogonal frame in the desired way. It is left to show that at the end of the inductive construction, we have $b_i^\perp = \sum_{j \neq i} b_j$. At every step of the procedure, we already have

$$b_i \leq \Big(\sum_{j<i} b_j \Big)^\perp = \prod_{j<i} b_j^\perp \leq b_j^\perp \quad \text{for } j < i.$$

In particular, for a arbitrary index $i < n + k$, we have $b_i \leq b_j^\perp$ for all $j \neq i$. Consequently, for fixed $i < n + k$, we have

$$b_i = b_i^\perp \cdot d = b_i^\perp \cdot \Big(\sum_{j<n+k} b_j \Big)$$

$$= \quad b_i^\perp \cdot \Big(b_i + \sum_{j\neq i} b_j \Big) = b_i^\perp \cdot b_i + \sum_{j\neq i} b_j$$

$$= \quad \sum_{j\neq i} b_j.$$

◁

1.5 Regular Rings and Lattices

In this section, we recall the connection between regular (∗-regular) rings and complemented lattices (modular ortholattices). Since all the results are folklore, most proofs are omitted or only roughly sketched.

Theorem 1.5.1. *A ring with unit is regular if and only if the set of all its principal right ideals is a complemented modular lattice.*

If R does not contain a unit, the equivalence holds for complemented *replaced by* relatively complemented.

Remark 1.5.1. For a ring R, we denote the set of all its principal right ideals by $\overline{L}(R_R)$. Likewise, we denote the set of all principal left ideals of R by $\overline{L}(_R R)$. Of course, both are partially ordered sets, without any additional assumption on the ring. In this thesis, we will deal with these partially ordered sets of principal one-sided ideals mainly in the case that R is a regular ring. Thence, we consider $\overline{L}(R_R)$ and $\overline{L}(_R R)$ to be modular lattices.

Lemma 1.5.2. *A ring with (without) unit is* ∗-*regular if and only if* $\overline{L}(R_R)$ *is a (sectional) MOL.*

▷ Proof. Folklore. If R is ∗-regular, every principal right ideal is generated by a projection. The orthogonality on $\overline{L}(R_R)$ is given by

$$aR \perp bR \ \Leftrightarrow \ b^*a = 0.$$

If R contains a unit, then the orthocomplement of eR, e a projection in R, is given by $(1 - e)R$. ◁

Proposition 1.5.3. *If* R *is regular, then the lattices* $\overline{L}(R_R)$ *and* $\overline{L}(_RR)$ *are anti-isomorphic.*
  If R *is* ∗-*regular,* $\overline{L}(R_R)$ *and* $\overline{L}(_RR)$ *are isomorphic.*

▷ Proof. See [Mic03], [Skor64] and [Mae58]. ◁

Definition 1.5.4. Height of a ∗-**regular ring**
The *height* $h(R)$ of a ∗-regular ring R is defined by the height $h(\overline{L}(R_R))$ of its principal ideal lattice $\overline{L}(R_R)$.

Lemma 1.5.5. *If* R *is a simple* ∗-*regular ring with (without) unit, then* $\overline{L}(R_R)$ *is a simple (sectional) MOL.*

▷ Proof. See [HR99, Theorem 2.5]. If we denote the set of all two-sided ideals of R by $\mathcal{A}$ and the set of all neutral ideals of $\overline{L}(R_R)$ by $\mathcal{B}$, then the 1-1-correspondence between two-sided ideals of R and neutral ideals of $\overline{L}(R_R)$ is given by the following maps.

$$F : \mathcal{A} \to \mathcal{B} \quad I \mapsto \{aR : a \in I\}$$

$$G : \mathcal{B} \to \mathcal{A} \quad \mathcal{I} \mapsto \{a \in R : aR \in \mathcal{I}\}$$

◁

Corollary 1.5.6. *Let R be a simple $*$-regular ring with unit and $h(R) \geq n$.*

Then the MOL $\overline{L}(R_R)$ contains a stable orthogonal (n, k)-frame.

Lemma 1.5.7. *In a simple MOL, each non-trivial interval $[0, a]$ is simple.*

▷ Proof. [Jón60, Lemma 2.2]. ◁

Lemma 1.5.8. *If R is a $*$-regular ring and e a projection in R, then the set eRe is a $*$-regular subring (with unit e) of R.*

▷ Proof. Clearly, eRe is closed under addition and multiplication. Since $e^* = e$, the involution on R restricts to an involution on eRe and is positive. For regularity, consider $x \in eRe$. Then $ex = xe = exe = x$. As an element of R, x has a quasi-inverse $y \in R$. But then we have

$$x = xyx = (xe)y(ex) = x(eye)x,$$

so $eye \in eRe$ is a quasi-inverse of x in eRe. ◁

Remark 1.5.2. For a projection e in a $*$-regular ring R, we sometimes use the abbreviation R_e for the $*$-regular subring eRe.

The following is due to Jónsson [Jón60, Lemma 8.2].

Lemma 1.5.9. *Let R be a $*$-regular ring and e a projection in R.*

Then the lattice $\overline{L}((R_e)_{R_e})$ of all principal right ideals of R_e is isomorphic to the section $[0, eR] \leq \overline{L}(R_R)$.

Lemma 1.5.10. *If R is a simple $*$-regular ring and e a projection in R, then R_e is a simple $*$-regular ring with unit.*

▷ **Proof.** For simplicity, let $A \neq \{0\}$ be an ideal in eRe. Since A does not vanish, the ideal $\langle A \rangle_R$ generated by A in R does not vanish, too. Hence, in the simple ring R, we have $\langle A \rangle_R = R$. Therefore, each element $x \in R$ can be written as

$$x = \sum r_i a_i s_i + \sum r_j a_j + \sum a_k s_k + \sum a_l$$

with $r_i, s_i, r_j, s_k \in R$ and $a_i, a_j, a_k, a_l \in A$. Multiplying by e from the left and the right and using the fact that for each $a \in A$, we have $ea = ae = eae = a$, we obtain

$$
\begin{aligned}
exe &= \sum er_i e a_i e s_i e + \sum er_j e a_j e \\
&\quad + \sum e a_k e s_k e + \sum e a_l e \\
&= \sum (er_i e) a_i (es_i e) + \sum (er_j e) a_j \\
&\quad + \sum a_k (es_k e) + \sum a_l.
\end{aligned}
$$

Consequently, $A = eRe$. ◁

Lemma 1.5.11. *Let R be a $*$-regular ring.*
  Then for each $x \in R$, there exists a projection $e_x \in R$ such that $e_x x e_x = x$.

▷ **Proof.** Let p_x be the projection that generates the left ideal generated by x, and let q_x be the projection that generates the right ideal generated by x. Take e_x to be the supremum of p_x and q_x in the lattice of all projections of R. ◁

Corollary 1.5.12. *If R is a simple $*$-regular ring and e a projection in R, then the lattice $\overline{L}((R_e)_{R_e})$ of principal right ideals of R_e contains a stable orthogonal frame.*

▷ **Proof.** This follows directly from the previous results. Of course, the format of the frame depends on the height of R_e. ◁

1.6 Representations

In this section, we introduce our notions of a *representation* of an involutive ring and a MOL, respectively. We begin the section with some basics.

Vector Spaces and Forms

We assume familiarity with the following objects: *Involutive skew fields, one-sided vector spaces over involutive skew fields, forms on (one-sided) vector spaces.*

Throughout the thesis, D will denote a skew field (in most cases an involutive skew field), V_D a right vector space over D, and ϕ a non-degenerated sesqui-linear form on V_D (if not explicitly stated otherwise).

In the case of an involutive skew field $(D,^*)$, we allow the involution to be trivial, i.e., we allow $^* : D \to D$ to coincide with $id_D : D \to D$. In that case, D is commutative.

We consider forms with the following properties:

1. We call ϕ a $*$-Hermitian form if $\phi(y, x) = \phi(x, y)^*$ for all $x, y \in V$.

2. We call ϕ a *symplectic form* if $\phi(x,x) = 0$ for all $x \in V$.

3. We call ϕ a *scalar product* if ϕ is an anisotropic $*$-Hermitian form.

Lemma 1.6.1. *The properties of a form to be $*$-Hermitian or to be symplectic do not exclude each other.*

▷ Proof. Let $D = \mathbb{F}_2$ (or any other field of characteristic two with trivial involution).

Consider a two-dimensional vector space over D with basis $\{a, b\}$ and define a (sesqui-linear) form ϕ via

$$\phi(a,a) = 0 = \phi(b,b) \quad \text{and} \quad \phi(a,b) = 1 = \phi(b,a).$$

Then ϕ satisfies the conditions of a $*$-Hermitian form; in particular

$$\phi(v,w) = (\phi(w,v))^* = \phi(w,v) = -\phi(w,v),$$

since $*$ is the identity and $+1 = -1$. ◁

Lemma 1.6.2. *Let ϕ be a symplectic form on V_D.*

Then D is a commutative field and the involution is the identity on D.

▷ Proof. Let v, w be two linearly independent vectors in V_D with $\phi(v,w) \neq 0$. Using linearity in the second argument, we can scale w such that $\phi(v,w) = 1$. Since ϕ is symplectic, we have $\phi(w,v) = -1$.

Let $s \in D$ be arbitrary. We consider $\phi(v + ws, v + ws)$.

Using sesqui-linearity and isotropy, we get

$$
\begin{aligned}
0 &= \phi(v + ws, v + ws) \\
&= \phi(v, v) + \phi(v, ws) + \phi(ws, v) + \phi(ws, ws) \\
&= \phi(v, w)s + s^*\phi(w, v) = s - s^*,
\end{aligned}
$$

that is, $s^* = s$.

Let $r, s \in D$ be arbitrary scalars. We consider the term $\phi(vr + ws, vr + ws)$. Using sesqui-linearity and isotropy, we get

$$
0 = r\phi(v, w)s + s\phi(w, v)r = rs - sr,
$$

so D is commutative. ◁

Corollary 1.6.3. *The example described above is the typical case of a form which is both $*$-Hermitian and symplectic, that is, if V_D is a vector space over an involutive skew field D and ϕ a non-degenerated form on V_D which is both $*$-Hermitian and symplectic, then D is already a commutative field with $\mathrm{char}(D) = 2$ and the involution is the identity on D.* [4]

▷ **Proof.** Since ϕ is symplectic, D is a commutative field and the involution is the identity. To show that $\mathrm{char}(D) = 2$, consider two linearly independent elements $v, w \in V_D$ such that $\phi(v, w) = 1$. Clearly,

$$
\phi(v + w, v + w) = 0,
$$

---

[4] In part., $\dim(V) \geq 2$, since ϕ is symplectic and non-degenerated.

since ϕ is symplectic. Furthermore,

$$\begin{aligned}\phi(v+w,v+w) &= \phi(v,v)+\phi(v,w)+\phi(w,v)+\phi(w,w)\\ &= \phi(v,w)+\phi(w,v),\end{aligned}$$

that is, $\phi(v,w) = -\phi(w,v)$. But since ϕ is $*$-Hermitian and the involution $* : D \to D$ is the identity on D, we have

$$\phi(v,w) = \big(\phi(w,v)\big)^* = \phi(w,v).$$

Therefore, we have $-\phi(w,v) = \phi(w,v)$. Consequently, $1 = -1$, which is to say that $char(D) = 2$. ◁

Definition 1.6.4. Hyperbolic plane
A two-dimensional non-degenerated subspace E of V will be called a *hyperbolic plane* if E has a basis of two isotropic vectors a, b.

If ϕ is $*$-Hermitian, we can normalise a, b such that $\phi(a,b) = 1 = \phi(b,a)$. If ϕ is symplectic, we can normalise a, b such that $\phi(a,b) = 1 = -\phi(b,a)$.

Definition 1.6.5. Metabolic plane
A two-dimensional non-degenerated subspace E of V with a $*$-Hermitian form will be called a *metabolic plane* if E has a basis $\{a, b\}$ such that a is isotropic.
Usually, we normalise the isotropic vector a such that $\phi(a,b) = 1 = \phi(b,a)$.

Observation 1.6.1. If ϕ is a symplectic form, clearly every metabolic plane is hyperbolic (since each vector is isotropic). If ϕ is $*$-Hermitian and $char(D) \neq 2$, each metabolic plane E is also hyperbolic, since we can construct two isotropic generating vectors of E.

Remark 1.6.1. Suprabolic plane Sometimes, we consider ∗-Hermitian forms and symplectic forms simultaneously. We then speak of *suprabolic planes*, meaning the appropriate of the above notions for each situation.

Representations of Rings

Definition 1.6.6. Linear representation
As in [Nie03], a *linear representation* of a regular involutive ring R is a tuple

$$\sigma = (D, V_D, \phi, \rho),$$

where D is an (involutive) skew field, V_D a right vector space over D, ϕ a non-degenerated sesqui-linear form on V_D, and

$$\rho : R \to End(V_D)$$

a ring homomorphism such that

$$\forall r \in R. \qquad \rho(r^*) = \rho(r)^{*\phi}.$$

We call σ a *linear symplectic*, *linear ∗-Hermitian*, or *linear positive representation* of R if the form ϕ is a symplectic form, a ∗-Hermitian form, or a scalar product, respectively. Of course, for *∗-regular* rings the notion of a linear *positive* representation is the appropriate one.

If the morphism $\rho : R \to End(V_D)$ is injective, we call σ a *faithful* representation.

Remark 1.6.2. For short, we speak of the representation $\rho : R \to End(V_D, \langle \cdot, \cdot \rangle)$ of R. If the ring R contains a unit 1_R, a representation $\rho : R \to End(V_D, \langle \cdot, \cdot \rangle)$ should map 1_R to id_V.

A more general notion of representability is the following.

Definition 1.6.7. Generalised representation
Let R be an involutive ring, I an arbitrary non-empty index set and σ a tuple

$$\sigma = (I, \{D_i\}_{i \in I}, \{V_i\}_{i \in I}, \{\phi_i\}_{i \in I}, \rho)$$

consisting of an indexed family of (involutive) skew fields, an indexed family of vector spaces, and an indexed family of non-degenerated forms such that for each $i \in I$, V_i is a right vector space over D_i with a symplectic, $*$-Hermitian, or positive $*$-Hermitian form ϕ_i, and a map

$$\rho : R \to \prod_{i \in I} End(V_{iD_i}).$$

If ρ is a $*$-ring morphism, i.e., for all $r \in R$ and all $i \in I$ the condition

$$\pi_i(\rho(r^*)) = \big(\pi_i(\rho(r))\big)^{*\phi_i}$$

holds, we call σ a *generalised representation* of R, a *g-representation* for short.

Remark 1.6.3. We call a g-representation σ *faithful* if the morphism ρ is injective. Note that we can only speak of a *symplectic, $*$-Hermitian* or *positive* g-representation if all forms ϕ_i are of the same type.

Remark 1.6.4. Note that the properties of a structure to be a *(faithful linear) representation of a ring* can be expressed in first-order logic (see [Mic03], p.34). Likewise,

the property of a ring to have a *(faithful) linear representation* or a *(faithful) g-representation* can be expressed
in first-order logic (see [Nie03, Proposition 6.2.1]).

Remark 1.6.5. Note that the notion of a representation makes sense for a ring R that is neither regular nor
equipped with an involution. For an arbitrary ring, a linear representation is just a ring morphism in the endomorphism ring of a vector space. We will use the accor-
dant terminology, that is, a ring that has a faithful linear
representation is just isomorphic as a ring to a subring
of an endomorphism ring of a vector space.

Representations of MOLs

Definition 1.6.8. Representation of a MOL
A *representation* of a (sectional) MOL L consists of a
tuple $\varsigma = (D, V_D, \langle \cdot, \cdot \rangle, \iota)$ with $D, V_D, \langle \cdot, \cdot \rangle$, as above and
a morphism $\iota : L \to L(V_D, \langle \cdot, \cdot \rangle)$ of bounded lattices (lat-
tices with bottom) between L and the subspace lattice of
V_D such that the (sectional) orthocomplementation on L
corresponds to the (sectional) orthocomplementation on
V_D given by the scalar product, that is, for all $x \in L$, we
have $\iota(x') = \iota(x)^\perp$ if L is a MOL ($\iota(x^b) = \iota(x)^\perp \cap \iota(b)$ if
L is a sectional MOL).

   We call a representation ς *faithful* if the morphism ι is
injective.

Remark 1.6.6. For short, we speak of the representa-
tion $\iota : L \to L(V_D, \langle \cdot, \cdot \rangle)$ of L. If the lattice L contains
a unit 1_L, a representation $\iota : L \to L(V_D, \langle \cdot, \cdot \rangle)$ should
map 1_L to V.

Theorem 1.6.9. *Let* $L = \overline{L}(R_R)$ *be a MOL coordinatised by some ∗-regular ring* R. *Assume that* L *has height at least 3.*

Then L *is faithfully representable, that is, there exists an injective MOL-embedding*

$$\iota : L \hookrightarrow L(V_D)$$

where V_D *is a vector space over a skew field* D *equipped with a scalar product* $\langle .,. \rangle$.

▷ Proof. [Herr] ◁

∗-Regular Rings and Representability

In this section, we recall some results presented in [Mic03]. We will make use of them in Chapter 4 in order to reduce the question of representability to a certain class of rings.

Micol proves the following.

Theorem 1.6.10. *A primitive ∗-regular ring with minimal one-sided ideal has a faithful linear positive representation.*

▷ Proof. See [Mic03, Theorem 3.3]. ◁

Remark 1.6.7. In the case of a ∗-regular ring, it is enough to require the existence of a minimal one-sided ideal, without insisting on left or right ideals.

Corollary 1.6.11. *A subdirectly irreducible atomic ∗-regular ring has a faithful linear positive representation.*

▷ Proof. See [Mic03, Corollary 3.4]. ◁

In the light of these results, we have the following.

Lemma 1.6.12. *A subdirectly irreducible Artinian ∗-regular ring has a faithful linear positive representation.*

In particular, a subdirectly irreducible ∗-regular ring with $h(R) \leq 2$ admits a faithful linear positive representation.

1.7 Universal Algebra

We assume familiarity with the following objects: *Algebra* or *algebraic structure* (of signature $\mathcal{L}$ or type $\mathcal{F}$), first-order algebraic structure, subdirect product, subdirect embedding, ultraproduct.

Subdirect Products

Theorem 1.7.1. *An algebra A is subdirectly irreducible iff A is trivial or there exists a minimal non-trivial congruence. In the latter case, this congruence is a principal congruence given by*

$$\bigcap_{\theta \in Con(\mathbf{A}) \setminus \{\Delta\}} \theta$$

▷ Proof. [UA00, Theorem II.8.4]. ◁

Theorem 1.7.2. *Every algebra $\mathbf{A}$ is isomorphic to a subdirect product of subdirectly irreducible algebras (which are homomorphic images of $\mathbf{A}$).*

▷ Proof. [UA00, Theorem II.8.6]. ◁

Ultraproducts

Theorem 1.7.3. *(Łoś)*
Let $\{\mathbf{A}_i\}_{i \in I}$ be a family of algebras of the same type, $\mathcal{U}$ an ultrafilter on I, and $\mathcal{P}$ a property which can be expressed in first-order logic. If $\mathcal{P}$ holds in every algebra $\mathbf{A}_i$, $i \in I$, then $\mathcal{P}$ holds in the ultraproduct $\prod_{i \in I} \mathbf{A}_i / \mathcal{U}$.

▷ Proof. [UA00, Theorem V.2.9]. ◁

Theorem 1.7.4. *(Malcev)*
Every first-order structure $\mathbf{A}$ can be embedded in an ultraproduct of its finitely generated substructures.

▷ Proof. [UA00, Theorem V.2.14]. ◁

Corollary 1.7.5. *Let R be a regular involutive ring. Then R can be embedded in an ultraproduct of its finitely generated regular involutive subrings R_i. In particular, each R_i is at most countable.*

2 On Representations of Regular Involutive Rings

This thesis represents the outcome of three years research.
The results are presented in seven short chapters.
Hence, you can plan a week's daytime reading
and even take the weekend off
– provided you leave the appendix
and the open questions for another time.

— Adapted from
the first edition of REPRESENTATIONS
AND CHARACTERS OF GROUPS
by Gordon James and Martin Liebeck

In this chapter, we generalise and complete results presented in [Mic03] and [Nie03].

In particular, we consider regular involutive rings that admit a faithful linear symplectic representation. On one hand, we will show that each countable such ring has a representation in a vector space of at most countable dimension. On the other hand, we will show that, if a regular involutive ring has a faithful linear symplectic representation in a vector space of countable dimension, then the matrix representations of its elements row finite with respect to an appropriate basis.

We finish the chapter with questions of finiteness of rings of operators. To be more specific, we examine certain operators x, y acting on vector spaces of countable dimension such that $xy = 1 \neq yx$. Under some assumptions with regard to the given basis and the characteristic of the underlying skew field, we are able to show that x and y cannot be contained in a regular involutive subring of the full endomorphism ring.

2.1 Countable Rings

In this section, we deal with faithful linear representations of countable regular involutive rings.

Linear ∗-Hermitian Representations

In [Mic03] and [Nie03], the following results have been shown.

Theorem 2.1.1. *Let R be a ∗-regular ring with a faithful linear positive ∗-Hermitian representation.*

If R is countable, then there exists a representation subspace of at most countable dimension.

▷ Proof. See [Mic03, Theorem 3.5] or [Nie03, Theorem 6.5.2]. ◁

Theorem 2.1.2. *Let R be a regular involutive ring with a faithful linear ∗-Hermitian representation.*

If R is countable, there exists a representation subspace of at most countable dimension.

▷ Proof. See [Nie03, Theorem 6.6.1] ◁

In the following section, we want to prove an analogous result for rings that admit a faithful linear symplectic representation.

Linear Symplectic Representations

Theorem 2.1.3. Subrepresentation of countable dimension

Let R be an at most countable regular involutive ring which has a faithful symplectic linear representation $\sigma = (D, W_D, \phi, \rho)$.

Then there exists a subspace U_D of W_D such that the dimension of U_D is at most countable and the restriction $\phi_0 := \phi_{|U_D}$ of ϕ to U_D is non-degenerated. Furthermore, there exists an embedding $\rho_0 : R \hookrightarrow End(U_D)$. Therefore, the tuple

$$\sigma_0 = (D, U_D, \phi_0, \rho_0)$$

is a faithful symplectic linear representation of R in the at most countably dimensional subspace U_D of the representation space.

▶ **Proof.** Consider $R^\times$, the set of all elements different from zero, and enumerate these elements with the natural numbers $\mathbb{N}$. We start with $r_0 := 1 = 1_R$. Then we consider the countable list

$$\big(\rho(r_0), \rho(r_1), \dots\big)^t.$$

Since σ is a faithful representation, we know that for each element $r_i \in R^\times$ there exists a vector $v_i \in W_D$ such that

$\rho(r_i)v_i \neq 0$. Thus, we can consider another countable list

$$(v_0, \widetilde{e}_0, v_1, \widetilde{e}_1, \dots),$$

where $\widetilde{e}_i$ is a wild card for the basis vector e_i which we want to define in the following process.

We use these two countable lists to define the following (countable) table.

	v_0	$\widetilde{e}_0$	v_1	$\widetilde{e}_1$	$\dots$
$\rho(r_0)$					
$\rho(r_1)$					
$\rho(r_2)$					
$\dots$					

In the following, we want

1. to fill out the cells of the table, and

2. to define the basis vectors e_i which will constitute the basis for the representation subspace.

We run through the table following a diagonal pattern, starting in the upper left corner. We indicate the pattern by numbers and then arrows.

	v_0	$\widetilde{e}_0$ (2)	v_1	$\widetilde{e}_1$(8)	$\dots$
$\rho(r_0)$	(1)	(4)	(7)	↗	$\dots$
$\rho(r_1)$	(3)	(6)	↗	$\dots$	
$\rho(r_2)$	(5)	↗	$\dots$		
$\vdots$	↗	$\dots$			

The cells Consider a cell of the table. This cell is determined by a line and a column, i.e. by a linear map $\rho(r_i)$

(at the left border of the table) and a vector (at the top border of the table), where the vector is either one of the chosen vectors v_i or an already defined basis vector e_i (notice that, due to our pattern, we reach the top of a column with a wild card character before we have to fill in any entry in that column). The entry of the cell is then given by the action of the corresponding linear map $\rho(r_i)$ on the vector v_i or e_i.

The basis vectors The first basis vector will be initialised with v_0: We set $e_0 := v_0$.

For the second basis vector, we choose a vector w which is not orthogonal to v_0. Such a vector exists since the form ϕ is non-degenerated. We scale such that $w' := \lambda w$ satisfies $\phi(e_0, w') = 1$; so e_0, w' constitute a hyperbolic basis of the two-dimensional subspace spanned by e_0 and w'. We set $e_1 := w'$.

Consider we reach a wild card character $\widetilde{e}_i$ for a basis vector which is not yet defined. Then we are going to choose vectors e_i and e_{i+1} that will be incorporated into our basis.

There might occur two situations: *Either* all entries $\rho(r_j)e_k, \rho(r_j)v_k$ calculated so far are contained in the span of the already defined basis vectors $e_1, \ldots, e_{i-1}$ *or* there exists an entry $\rho(r_j)e_k, \rho(r_j)v_k$ in the already calculated part of the table which is not contained in the span of the already defined basis vectors.

In the first case, we cross out the whole column and re-index the following wild card characters accordingly (i.e. decrement their indices). Then we continue with our process.

In the second case, we have to define the corresponding basis vector. But we have to take care that the subspace will be non-degenerated. So, let v be the first entry in the already calculated part of the table which is not contained in the span of the already defined basis vectors. Then consider the subspace S generated by the already defined basis vectors $e_0, \ldots, e_{i-1}$, that is,

$$S_{i-1} := span(e_0, \ldots, e_{i-1}).$$

Since W is non-degenerated and $dim(S_{i-1}) < \infty$, we get a decomposition of W into

$$W_D = S_{i-1} \oplus S_{i-1}^\perp.$$

Since $v \notin S_{i-1}$, we can represent v as

$$v = s + s', \quad \text{where } s \in S_{i-1} \text{ and } s' \in S_{i-1}^\perp.$$

We define $e_i := s'$. But now we have to make sure that our process does not yield a degenerated subspace. To do so, we we have to define the next basis vector on the spot (so, the presented pattern is just the general scheme). Since the form ϕ is non-degenerated and $s' \in S_{i-1}^\perp$, there exists a vector $u \in S_{i-1}^\perp$ such that $\phi(e_i, u) \neq 0$.

By scaling u, we get a vector u' such that $\phi(e_i, u') = 1$. We set $e_{i+1} := u'$.

Following this pattern, we get a set $\{e_i : i \in I\}$, where $I \subseteq \mathbb{N}$, which constitutes a basis of a subspace U_D of W_D such that the restriction of ϕ to U_D is non-degenerated. Furthermore, the construction yields a hyperbolic basis of U.

Recalling that to each $r_i \in R^\times$ we have chosen v_i with

$\rho(r_i)v_i \neq 0$ and each such v_i is an element of U, we have that $\rho : R \to End(U_D)$ is injective, so the constructed representation is faithful. ◀

2.2 Matrix Representations

Linear *-Hermitian Representations

In [Mic03] and [Nie03], the following results have been shown.

Theorem 2.2.1. *Let R be a *-regular ring with linear positive *-Hermitian representation σ. Assume that the representation space V_D has an orthogonal basis α with respect to the scalar product ϕ.*

Then the matrix representation $[\rho(r)]^\alpha$ of each element $r \in R$ is a countable small matrix.

▷ Proof. See [Mic03, Theorem 2.10]. ◁

Theorem 2.2.2. *Let R be a regular involutive ring with linear *-Hermitian representation σ. Assume that the representation space V_D has an orthogonal decomposition into non-degenerated lines and planes. Let Λ be the compatible basis with respect to the given decomposition.*

Then the matrix representation $[\rho(r)]^\Lambda$ of each element $r \in R$ is a countable small matrix.

▷ Proof. See [Nie03, Corollary 6.6.3]. ◁

Remark 2.2.1. For the notion of a *compatible basis* with respect to a given decomposition, see either [Nie03] or Section 2.2, Definition 2.2.4.

In the following section, we want to prove an analogous result for rings that admit a faithful linear symplectic representation.

Linear Symplectic Representations

This section starts with the generalised Gram-Schmidt-orthogonalisation for a symplectic vector space of countable dimension. Then we will introduce the concept of a compatible basis for an orthogonal decomposition. Given that, we want to prove certain relations between the entries of the matrix representation of a continuous operator and the matrix representation of its adjoint with respect to the compatible basis. These relations enable us to prove the desired result of row finiteness of the matrix representations of continuous operators with respect to the compatible basis.

Gram-Schmidt-Orthogonalisation

In the spirit of H. Gross (see [Gross79, Chapter 2]), we want to prove the following.

Theorem 2.2.3. *Let D be a field, W_D a right vector space over D and ϕ a symplectic form on W_D.*
If W_D has at most countable dimension, then there exists a decomposition of W_D into pairwise orthogonal non-degenerated two-dimensional subspaces of W_D.

▶ **Proof.** We start with a basis $\beta = \{e_i\}_{i \in I}$ where the index set is a finite or infinite subset of $\mathbb{N}$. Without loss of generality, we can assume $I = \mathbb{N}$.

Our aim is to construct a family of two-dimensional, pairwise orthogonal, non-degenerated subspaces F_k of W_D such that the following hold.

$$\forall m \in I. \qquad i < m \Rightarrow e_i \in F_0 \oplus \ldots \oplus F_m$$

We follow a generalisation of the process of the Gram-Schmidt-orthogonalisation. In particular, we prove the existence of the desired family by exhibiting an inductive procedure.

We start with the first basis vector e_0. Since the form ϕ is non-degenerated, there exists a vector $w \in V$ such that $\phi(e_0, w) \neq 0$. We define $F_0 := span(e_0, w)$.

Now assume that we have accomplished the construction of a finite sequence $F_0, F_1, \ldots, F_m$ of pairwise orthogonal non-degenerated planes F_i with the desired property. We set

$$S := F_0 \oplus F_1 \oplus \ldots \oplus F_m.$$

Then S is a non-degenerated finite-dimensional subspace of W_D. We have $codim(S^\perp) = dim(S)$ and, since the form ϕ is non-degenerated, $S \cap S^\perp = \{0\}$.

Using the isomorphism theorems, we get

$$(S + S^\perp)/S^\perp \cong S/(S \cap S^\perp) \cong S.$$

Furthermore, we have the inclusion

$$(S + S^\perp)/S^\perp \subseteq W/S^\perp.$$

Since we have

$$dim(W/S^{\perp}) = codim(S^{\perp}) = dim(S)$$
$$= dim((S + S^{\perp})/S^{\perp}) < \infty,$$

we get $W/S^{\perp} = (S + S^{\perp})/S^{\perp}$ and so $W = S + S^{\perp}$. Since the intersection of S and $S^{\perp}$ is trivial, we have $W = S \oplus S^{\perp}$. Now we distinguish two cases.

Case 1: $S^{\perp} = \{0\}$. Then we set $J = \{0, \ldots, m\}$ and are done.

Case 2: $S^{\perp} \neq \{0\}$. Then $m + 1 \in I$ and $e_{m+1} \notin S$, i.e.,

$$e_{m+1} = s + s' \qquad s \in S, \ s' \in S^{\perp}, \ s' \neq 0$$

Since ϕ is non-degenerated, there is a vector $x \in W_D$ such that $\phi(x, s') \neq 0$. We wish to construct an element $x' \in S^{\perp}$ such that $\phi(x', s') \neq 0$.

So, assume that $x \notin S^{\perp}$. Since $W = S \oplus S^{\perp}$, we can write x as $x = a + b$ with $a \in S^{\perp}, b \in S$ with $b \neq 0$. Therefore, we define $x' := a$.

We notice that $a \neq 0$, or else, we would have $x = b \in S$. Since $s' \in S^{\perp}$, this would imply that $\phi(x, s') = 0$.

We set $F_{m+1} := span(e_{m+1}, x')$. Thus, we have enlarged our set of pairwise orthogonal non-degenerated planes by one plane. If W_D is finite-dimensional, this process terminates. If the dimension of W_D is countably infinite, we have to rely on the axiom of dependent choice to get a countably infinite sequence of planes with the desired properties. ◄

Hyperbolic Decomposition and the Compatible Basis

Definition 2.2.4. Compatible basis

Let W_D be a right vector space over a field D, ϕ a non-degenerated symplectic form on W_D. Assume that W_D is decomposed into a direct sum of pairwise orthogonal hyperbolic planes, that is, $W = \bigoplus_{i \in I} F_i$. To each plane F_i, denote the hyperbolic basis by $\{a_i, b_i\}$.

Then we introduce the compatible basis Λ of W_D in the usual way. For the new index set. we take two copies A, B of the index set I and define the index set $\mathcal{I}$ to be the disjoint union of A and B. We denote the copy of i in A (B) by i_α (i_β). In particular, we have

$$F_i = span(a_{i_\alpha}, b_{i_\beta}).$$

If $j = i_\alpha \in \mathcal{I}$, we write $i_\beta = j_\beta$ and if $j = i_\beta \in \mathcal{I}$, we write $i_\alpha = j_\alpha$. Now we define the compatible basis $\Lambda := \{e_i : i \in \mathcal{I}\}$ as follows:

1. For $i \in A$, we set $e_i := a_i$.

2. For $i \in B$, we set $e_i := b_i$.

With this settings, the following relations hold.

1. $\forall i \in \mathcal{I}.$ $\phi(e_i, e_i) = 0.$

2. $\forall i, j \in A.$ $\phi(e_i, e_j) = 0.$

3. $\forall i, j \in B.$ $\phi(e_i, e_j) = 0.$

4. $\forall i \in A, j \in B, i_\beta = j.$ $\phi(e_i, e_j) = 1.$ [1]

5. $\forall i \in A, j \in B, i_\beta \neq j (rsp.\ i \neq j_\alpha).$ $\phi(e_i, e_j) = 0.$

---

[1] That is, $i = j_\alpha$ and $\phi(e_j, e_i) = -1$, respectively.

Adjoint Operators

Proposition 2.2.5. *Let W_D be a right vector space over a skew field D, ϕ a non-degenerated symplectic form on W_D and $r, s \in End(W_D)$ adjoint to each other with respect to ϕ. Assume that W_D is decomposed into a direct sum of pairwise orthogonal hyperbolic planes, that is, $W = \bigoplus_{i \in I} F_i$.*

Assume that $F_i = span(a_i, b_i)$ and let Λ be the induced basis of this decomposition.

Then the following relations hold.

    1. $\forall i, k \in A.$ $r_{i_\beta k} = -s_{k_\beta i}$

    2. $\forall i \in A, k \in B.$ $r_{i_\beta k} = s_{k_\alpha i}$

    3. $\forall i \in B, k \in A.$ $r_{i_\alpha k} = s_{k_\beta i}$

    4. $\forall i, k \in B.$ $- r_{i_\alpha k} = s_{k_\alpha i}$

▶ **Proof.** The proof follows the lines of the proofs presented in [Mic03] and [Nie03]. It is basically a brute force calculation.

Let $r, s \in End(W_D)$ be adjoint to each other. Then we have $\phi(v, rw) = \phi(sv, w)$ for all $v, w \in W_D$. We consider this relation for all possible pairs of basis vectors e_i, e_k of Λ. We recall that in general we have the following.

$$re_k = \sum_{n \in I} e_n r_{nk} = \sum_{a \in A} e_a r_{ak} + \sum_{b \in B} e_b r_{bk}$$

$$se_i = \sum_{m \in I} e_m s_{mi} = \sum_{a \in A} e_a s_{ai} + \sum_{b \in K} e_b s_{bi}$$

Now we proceed by case distinction.

Case 1 $i, k \in A$

$$
\begin{aligned}
\phi(e_i, re_k) &= \sum_{a \in A} \phi(e_i, e_a r_{ak}) + \sum_{b \in B} \phi(e_i, e_b r_{bk}) \\
&= \sum_{a \in A} \phi(e_i, e_a) r_{ak} + \sum_{b \in B} \phi(e_i, e_b) r_{bk} \\
&= 0 + \sum_{b \in B} \phi(e_i, e_b) r_{bk} \\
&= \phi(e_i, e_{i_\beta}) r_{i_\beta k} = r_{i_\beta k}
\end{aligned}
$$

On the other hand, we have the following.

$$
\begin{aligned}
\phi(se_i, e_k) &= \sum_{a \in A} \phi(e_a s_{ai}, e_k) + \sum_{b \in B} \phi(e_b s_{bi}, e_k) \\
&= \sum_{a \in A} \phi(e_a, e_k) s_{ai} + \sum_{b \in B} \phi(e_b, e_k) s_{bi} \\
&= 0 + \sum_{b \in B} \phi(e_b, e_k) s_{bi} \\
&= \phi(e_{k_\beta}, e_k) s_{k_\beta i} = -s_{k_\beta i}
\end{aligned}
$$

Therefore, we get the relation

$$
\forall i, k \in A. \qquad r_{i_\beta k} = -s_{k_\beta i}.
$$

Case 2 $i \in A, k \in B$

$$
\begin{aligned}
\phi(e_i, re_k) &= \sum_{a \in A} \phi(e_i, e_a r_{ak}) + \sum_{b \in B} \phi(e_i, e_b r_{bk}) \\
&= \sum_{a \in A} \phi(e_i, e_a) r_{ak} + \sum_{b \in B} \phi(e_i, e_b) r_{bk} \\
&= 0 + \sum_{b \in B} \phi(e_i, e_b) r_{bk} \\
&= \phi(e_i, e_{i_\beta}) r_{i_\beta k} = r_{i_\beta k}
\end{aligned}
$$

$$
\begin{aligned}
\phi(se_i, e_k) &= \sum_{a \in A} \phi(e_a s_{ai}, e_k) + \sum_{b \in B} \phi(e_b s_{bi}, e_k) \\
&= \sum_{a \in A} \phi(e_a, e_k) s_{ai} + \sum_{b \in B} \phi(e_b, e_k) s_{bi} \\
&= \sum_{a \in A} \phi(e_a, e_k) s_{ai} + 0 \\
&= \phi(e_{k_\alpha}, e_k) s_{k_\alpha i} = s_{k_\alpha i}
\end{aligned}
$$

Therefore,

$$
\forall i \in A, k \in B. \qquad r_{i_\beta k} = s_{k_\alpha i}.
$$

Case 3 $i \in B, k \in A$

$$
\begin{aligned}
\phi(e_i, re_k) &= \sum_{a \in A} \phi(e_i, e_a r_{ak}) + \sum_{b \in B} \phi(e_i, e_b r_{bk}) \\
&= \sum_{a \in A} \phi(e_i, e_a) r_{ak} + \sum_{b \in B} \phi(e_i, e_b) r_{bk} \\
&= \sum_{a \in A} \phi(e_i, e_a) r_{ak} + 0 \\
&= \phi(e_i, e_{i_\alpha}) r_{i_\alpha k} = -r_{i_\alpha k}
\end{aligned}
$$

$$
\begin{aligned}
\phi(se_i, e_k) &= \sum_{a \in A} \phi(e_a s_{ai}, e_k) + \sum_{b \in B} \phi(e_b s_{bi}, e_k) \\
&= \sum_{a \in A} \phi(e_a, e_k) s_{ai} + \sum_{b \in B} \phi(e_b, e_k) s_{bi} \\
&= 0 + \sum_{b \in B} \phi(e_b, e_k) s_{bi} \\
&= \phi(e_{k_\beta}, e_k) s_{k_\beta i} = -s_{k_\beta i}
\end{aligned}
$$

Therefore,

$$
\forall i \in B, k \in A. \qquad r_{i_\alpha k} = s_{k_\beta i}.
$$

Case 4 $i, k \in B$

$$
\begin{aligned}
\phi(e_i, re_k) &= \sum_{a \in A} \phi(e_i, e_a r_{ak}) + \sum_{b \in B} \phi(e_i, e_b r_{bk}) \\
&= \sum_{a \in A} \phi(e_i, e_a) r_{ak} + \sum_{b \in B} \phi(e_i, e_b) r_{bk} \\
&= \sum_{a \in A} \phi(e_i, e_a) r_{ak} + 0 \\
&= \phi(e_i, e_{i_\alpha}) r_{i_\alpha k} = -r_{i_\alpha k}
\end{aligned}
$$

$$
\begin{aligned}
\phi(se_i, e_k) &= \sum_{a \in A} \phi(e_a s_{ai}, e_k) + \sum_{b \in B} \phi(e_b s_{bi}, e_k) \\
&= \sum_{a \in A} \phi(e_a, e_k) s_{ai} + \sum_{b \in B} \phi(e_b, e_k) s_{bi} \\
&= \sum_{a \in A} \phi(e_a, e_k) s_{ai} + 0 \\
&= \phi(e_{k_\alpha}, e_k) s_{k_\alpha i} = s_{k_\alpha i}
\end{aligned}
$$

Therefore,

$$
\forall i, k \in B. \qquad -r_{i_\alpha k} = s_{k_\alpha i}.
$$

◀

Continuous Endomorphisms and Row Finiteness

Using the relations deduced in the previous section, we can prove a result similar to Theorem 2.2.1 and Theorem 2.2.2.

Corollary 2.2.6. *Let D be a field, W_D be a right vector space of countable dimension over D, ϕ a symplectic form on W_D and assume that W_D is decomposed in a direct sum of pairwise orthogonal non-degenerated hyperbolic planes, that is, $W = \bigoplus_{i \in I} H_i$. Let Λ be the compatible basis of this decomposition. Assume that $r, s \in End(W_D)$ are adjoint to each other with respect to ϕ.*

Then the matrix representations of r and s with respect to Λ are row finite.

▷ **Proof.** Consider a fixed index $p \in \mathcal{I}$. We want to show that there are only finitely many non-vanishing entries $r_{pk}, k \in \mathcal{I}$. Again, we proceed by case distinction.

Case 1 $p \in A$

Then we have to consider the following two situations.

1. $k \in A$: Then we have $r_{p_\beta k} = -s_{k_\beta p}$.

2. $k \in B$: Then we have $r_{p_\beta k} = s_{k_\alpha p}$.

Case 2 $p \in B$

Then we have to consider the following two situations.

1. $k \in A$: Then we have $r_{p_\alpha k} = s_{k_\beta p}$.

2. $k \in B$: Then we have $-r_{p_\alpha k} = s_{k_\alpha p}$.

In every possible case, the column finiteness of matrix representation of the endomorphism s with respect to Λ implies that there are only finitely many non-vanishing entries $r_{pk}, k \in \mathcal{I}$. Therefore, the matrix representation of r with respect to Λ is row finite. So we have shown that in the given situation, each endomorphism having an adjoint has a row and column finite matrix representation with respect to Λ. ◁

2.3 $End(V_D)$: Folklore

In the following section, we recall some well-known results about endomorphism rings of vector spaces.

Regularity vs. Continuity

Proposition 2.3.1. *Let V_D be a vector space of arbitrary dimension over a skew field D.*

Then the full endomorphism ring $End(V_D)$ is regular.

▷ Proof. [Mic03, Proposition 2.1]. ◁

Combining the results about symplectic, $*$-Hermitian and positive $*$-Hermitian representations, we get the following.

Proposition 2.3.2. *Let V_D be a vector space over D, ϕ a scalar product or a non-degenerated $*$-Hermitian or symplectic form on V_D. Assume that we have an orthogonal decomposition of V_D. Let β be the compatible basis of V_D with respect to that decomposition.*

Then the continuous endomorphisms in End(V_D) are exactly the endomorphisms whose matrix representation with respect to β is a countable small matrix.

Moreover, explicit formulas can be given for the relations of the entries of the matrix representations of two operators a, b which are adjoint to each other.

Especially, all endomorphisms having a finite matrix representation with respect to β are continuous and the set of the endomorphisms having a finite matrix representation with respect to β is closed under adjunction.

Using this result, we can generalise [Mic03, Corollary 2.11] in the following way.

Proposition 2.3.3. *Let V_D be a vector space over D and ϕ a non-degenerated form on V, where ϕ can be either symplectic, $*$-Hermitian, or positive $*$-Hermitian.*

Assume that we have an orthogonal decomposition of V_D with respect to ϕ and let β be the corresponding compatible basis of V_D.

Then the ring $Cont_\phi(V_D)$ of all continuous endomorphisms is regular iff V_D is finite-dimensional.

The proof is the very same as in [Mic03].

▶ **Proof.** Assume that V_D has finite dimension. Then every endomorphism of V_D is continuous. Proposition 2.3.1 proves regularity.

Assume that V_D has infinite dimension. We choose a countable subset of the compatible basis $\beta = \{e_i : i \in I\}$, i.e., we decompose the index set I into $I = \mathbb{N} \cup J$. We define the endomorphisms f, g by their actions on the basis vectors.

$$f(e_i) = \begin{cases} e_i & \text{if } i \in J \\ e_0 & \text{if } i = 0 \\ e_i + e_{i-1} & \text{if } i \in \mathbb{N} \setminus \{0\} \end{cases}$$

$$g(e_i) = \begin{cases} e_i & \text{if } i \in J \\ e_0 & \text{if } i = 0 \\ \sum_{k=0}^{i}(-1)^{k-1}e_k & \text{if } i \in \mathbb{N} \setminus \{0\} \end{cases}$$

Note that the matrix representation of f with respect to β is row and column finite, thus f is continuous. Since the matrix representation of g with respect to β is not row finite, g is not continuous. Because we have $(g \circ f)(e_i) = e_i = (f \circ g)(e_i)$ for every basis element $e_i \in \beta$, g is the inverse of f. Consequently, g is the only possible quasi-inverse for f in the ring $End(V_D)$ of all endomorphisms. Hence, f has no quasi-inverse in the ring $Cont_\phi(V_D)$ of all continuous endomorphisms. ◄

Subrings

It should be mentioned that it is not an easy task to identify regular (involutive) subrings of endomorphism rings of vector spaces; that is, to answer the question whether a given subring is regular, or to describe regular subrings of endomorphism rings. In this section, we give an easy example of a non-regular subring of an endomorphism ring.

Consider a nilpotent endomorphism $a \in End(V_D)$. Let R be the ring generated by a and 1. Then the chance that R is a regular ring is rather small.

Example 2.3.1. Choose $D := \mathbb{R}$ and let V_D be the two-dimensional vector space over D. Let $\alpha = \{e_0, e_1\}$ be the canonical basis. Assume that V is equipped with the standard scalar product; so, the involution on $End(V_D)$ with respect to ϕ is given by transposition of the matrix representations of endomorphisms with respect to α.

Consider the nilpotent linear operator $a : V \to V$ given by

$$a : V \to V \qquad e_0 \mapsto 0 \quad e_1 \mapsto e_0$$

The matrix representation of a with respect to α is given by

$$A := \begin{pmatrix} 0 & 1 \\ 0 & 0 \end{pmatrix}.$$

Consider the involutive ring $R := R(1, a)$ generated by a and 1. Then an element b of $End(V)$ belongs to R if and only if it has a matrix representation with respect to α of the form

$$B = \begin{pmatrix} a & b \\ c & a \end{pmatrix}, \qquad \text{with } a, b, c \in \mathbb{Z}.$$

R is an involutive subring of $End(V)$; but it is easy to see that R is not regular: Consider

$$X := \begin{pmatrix} 1 & 2 \\ 2 & 1 \end{pmatrix} \in R.$$

Then there exists no $Y \in R$ such that $YXY = X$: Since X is invertible in $End(V)$, the only possible quasi-inverse is the inverse of X. But X^{-1} has the following matrix representation with respect to α:

$$X^{-1} = -\frac{1}{3}\begin{pmatrix} 1 & -2 \\ -2 & 1 \end{pmatrix}.$$

In particular, X^{-1} does not lie in R.

2.4 $End(V_D)$: Finiteness of Subrings

Notation

Definition 2.4.1. Directly finite ring
A ring is called *directly finite* or *von Neumann-finite* or *Dedekind-finite* if the implication

$$xy = 1 \Rightarrow yx = 1$$

holds.

Definition 2.4.2. Finite ring
An involutive ring is called *finite* if the implication

$$xx^* = 1 \Rightarrow x^*x = 1$$

holds.

Shift Operators

For the subsequent section(s), we assume D to be an (involutive)skew field of infinite characteristic.

We will examine endomorphism rings of vector spaces of countable dimension, equipped with a non-degenerated form and some particular operators not satisfying the implication $xy = 1 \Rightarrow yx = 1$; we will show that they are not contained in a regular involutive subring of the full endomorphism ring.

Example 2.4.1. Shift operators on a positive space

Let V_D be a vector space over D, equipped with a scalar product ϕ. Assume that V has countable dimension and that $\beta = \{e_i : i \in \mathbb{N}\}$ is an orthogonal basis of V_D with respect to ϕ.

Let $a \in End(V)$ be the right shift operator on the basis β, defined by

$$a : V \to V \qquad e_i \mapsto e_{i+1}$$

Then the following holds:

1. The adjoint of a with respect to ϕ is the left shift operator a^* on β given by

$$a^* : V \to V \qquad e_0 \mapsto 0 \qquad e_i \mapsto e_{i-1} \qquad i \in \mathbb{N}\backslash\{0\}.$$

2. The operators a, a^* satisfy $a^*a = 1$, but $aa^* \neq 1$.

3. There is no regular involutive subring S of $End(V_D)$ containing a (and its adjoint a^*).

Of course, these statements are folklore. The fact that a, a^* are not contained in a $*$-regular subring of $End(V)$ already follows from the more general result of Ara and Menal that every $*$-regular ring is finite. Nevertheless, we will present an alternative proof, making use of the given structure, since the line of arguments will be used later during the consideration of vector spaces with more general forms.

▶ **Proof.** We consider the matrix representations of a and a^* with respect to β.

$$A := [a]^\beta = \begin{pmatrix} 0 & 0 & 0 & 0 & \dots \\ 1 & 0 & 0 & 0 & \dots \\ 0 & 1 & 0 & 0 & \dots \\ 0 & 0 & 1 & 0 & \dots \\ \vdots & \vdots & \vdots & \ddots & \dots \end{pmatrix}$$

$$A^* := [a^*]^\beta = \begin{pmatrix} 0 & 1 & 0 & 0 & \dots \\ 0 & 0 & 1 & 0 & \dots \\ 0 & 0 & 0 & 1 & \dots \\ \vdots & 0 & 0 & 0 & \ddots & \dots \end{pmatrix}$$

We define $x := 1 + a$. Then the matrix representation of x with respect to β is given by

$$X := [x]^\beta = \begin{pmatrix} 1 & 0 & 0 & 0 & \dots \\ 1 & 1 & 0 & 0 & \dots \\ 0 & 1 & 1 & 0 & \dots \\ 0 & 0 & 1 & 1 & \dots \\ \vdots & \vdots & \vdots & \ddots & \dots \end{pmatrix}.$$

Furthermore, each (regular involutive) subring of $End(V)$ which contains a has to contain x as well, since subring means subring with 1. We will show that there exists no continuous quasi-inverse of x in $End(V)$.

Following [Mic03], we can write down at least one

quasi-inverse y of x in $End(V)$, namely

$$Y := [y]^\beta = \begin{pmatrix} 0 & 1 & -1 & 1 & -1 & \cdots \\ 0 & 0 & 1 & -1 & 1 & -1 & \cdots \\ 0 & 0 & 0 & 1 & -1 & 1 & -1 & \cdots \\ \vdots & & \cdots \end{pmatrix}.$$

Taking a closer look at the matrix representations of y and x, we can see that y is not only a quasi-inverse of x, but even a left-inverse of x.

We have the following result.

Lemma 2.4.3. *Let R be a ring, $r, s \in R$ such that s is a left-inverse of r. Then each quasi-inverse of r is a left inverse of r.*

▷ **Proof.** Assume that t is a quasi-inverse of r; so $rtr = r$. Multiplying with s from the left, we get $s(rtr) = sr$; using associativity and $sr = 1$, we have $tr = 1$. ◁

Proposition 2.4.4. *Let z be a quasi-inverse (hence a left inverse) of $x = 1 + a$.*

Then the matrix representation of z with respect to β is not row finite. Consequently, there cannot exist an operator which is adjoint to z.

▶ **Proof.** We proceed by case distinction over the possible entries of $Z := [z]^\beta$. We begin with the entry z_{00} and distinguish the two cases $z_{00} \neq 0$ and $z_{00} = 0$.

Case 1: $z_{00} \neq 0$. This implies $z_{00} + z_{01} = 1$, which is equivalent to $z_{01} = 1 - z_{00}$. Now, we consider the cases $z_{00} = 1$ and $z_{00} \neq 1$ (but still $z_{00} \neq 0$).

Case 1.1: $z_{00} = 1$. This implies $z_{01} = 0$; now we have to examine the second row. Here, we distinguish the cases $z_{10} = 0$ and $z_{10} \neq 0$.

Assume that $z_{10} = 0$. Because the product of the second row of Z with the first column of X has to vanish, this implies $z_{11} = 0$. Considering the product of the second row of Z with the second column of X, we obtain that $z_{12} = 1$. Followed by an examination of the product of the second row of Z with the other columns of X, we see that Z has to have alternating entries of 1 and -1 in the second row.

Assume that $z_{10} \neq 0$. Since the product of the second row of Z and the first column of X has to vanish, this implies that $z_{11} = -z_{10}$. Now, we consider the product of the first row of Z and the second column of X. Since this has to equal 1, we get that $z_{12} = 2\lambda$. Considering further products of the second row of Z and the columns of X, we see that the second row of Z has infinitely many non-vanishing entries $\lambda, -\lambda, 2\lambda, -2\lambda, \ldots$.

Case 1.2: $z_{00} \neq 1$, $z_{00} \neq 0$. We need $z_{00} + z_{01} = 1$, so $z_{01} = 1 - z_{00} \neq 0$. Now, considering the product of the first row of Z with the other columns of X, we see that the first row of Z has infinitely many non-vanishing entries $\lambda, -\lambda, 2\lambda, -2\lambda, \ldots$.

Case 2: $z_{00} = 0$. It follows that $z_{01} = 1$. Consequently, all following entries of the first row of Z are, by turns, 1 or -1.

In every case, the matrix representation of Z with respect to β is not row-finite, hence, there exists no operator in $End(V)$ which is adjoint to z. ◄

The above argument shows that a, a^* cannot be contained in a $*$-regular subring of $End(V)$. ◄

Shift Operators on General Spaces   After the previous introductory example of shift operators on spaces equipped with a scalar product, we consider a more general situation. Let V_D a vector space of countably infinite dimension over D equipped with a non-degenerated $*$-Hermitian or symplectic form ϕ. Furthermore, assume that V is decomposed into a direct sum

$$V = \bigoplus_{i \in \mathbb{N}} F_i$$

of pairwise orthogonal non-degenerated planes F_i.

Let Λ be the compatible basis with respect to the given decomposition. We consider the right and the left shift operator on V, given by

$$a : V \to V \qquad e_i \mapsto \begin{cases} e_{i'} = e_{i_B} & \text{if } i \in A \\ e_{i'+1} = e_{i_A+1} & \text{if } i \in B \end{cases}$$

$$b : V \to V \qquad e_i \mapsto \begin{cases} 0 & \text{if } i = 0_A \\ e_{i'-1} = e_{i_B-1} & \text{if } i \in A, i \neq 0 \\ e_{i'} = e_{i_A} & \text{if } i \in B \end{cases}$$

We denote the matrix representation of a morphism $f : V \to V$ with respect to Λ in the following way: The first entry (of the first row) corresponds to (e_0, e_0), the second entry to (e_0, e_0'), the third to (e_0, e_1) and so on.

The matrix representations A, B of a, b with respect to Λ are then given by the same matrices as in the previous section.

Again, we consider the operator $x := 1 + a$. The argument presented above applies to this situation as well: We can write down the same left-inverse y in $End(V)$, use y to show that every quasi-inverse of x is a left-inverse of x and follow the same case distinction to prove that a left-inverse z of x cannot have a row-finite matrix representation with respect to Λ. Then the results of Section 2.2 yield that z has no adjoint operator. Therefore, a and b cannot be contained in a regular involutive subring of $End(V)$.

Generalised Shift Operators I: Powers

We consider the same situation as in the previous section. Again, we denote the right shift operator by a and the left shift operator by b. For some natural number $k \geq 2$, we define $f := a^k$ and $g := b^k$.

Proposition 2.4.5. *There exists no regular involutive subring of $End(V)$ containing the powers f and g of the usual shift operators a and b.*

▶ **Proof.** We define $x := 1 + f$. Then the matrix representation of x is given by

$$
X := [x]^\Lambda = \begin{pmatrix}
1 & 0 & 0 & 0 & \dots \\
0 & 1 & 0 & 0 & \dots \\
\vdots & 0 & 1 & 0 & \dots \\
0 & \vdots & 0 & 1 & \dots \\
1 & 0 & \vdots & \vdots & \dots \\
0 & 1 & \vdots & \vdots & \dots
\end{pmatrix},
$$

where the second non-vanishing entry in the first column has row index k.

Now, for an arbitrary left inverse of x, we can follow a similar case distinction as in the previous situation; we just have to consider different indices. For example, if $z_{00} = 0$, then z_{0k} has to be 1. But then we get infinitely many entries in the first row of Z: Alternating entries of 1 and -1 at positions k, $2k$, $3k$, etc. ◄

Generalised Shift Operators II

In the literature, another generalised shift operator can be found in [Lam01, Proposition 11.23]. This time, the shift depends on the index of the considered basis vector – the higher the index, the further the basis vector gets shifted. More exactly, on a vector space W with basis $\beta = \{b_i : i \in \mathbb{N}\}$, Lam defines the following operator via its following action on the basis β.

$$
f : W \to W \qquad f : a_i \mapsto a_{i^2+1}
$$

We consider such an operator in a setting as above, that is, we take f to be the linear operator with the above action on a compatible basis Λ.

Proposition 2.4.6. *There exists no involutive subring of $End(V_D)$ containing f.*

▶ **Proof.** The matrix representations of f and its adjoint $g := f^*$ with respect to Λ are given by

$$
F := [f]^\Lambda = \begin{pmatrix}
0 & 0 & 0 & 0 & \cdots \\
1 & 0 & 0 & 0 & \cdots \\
0 & 1 & 0 & 0 & \cdots \\
0 & 0 & 0 & 0 & \cdots \\
0 & 0 & 0 & 0 & \cdots \\
0 & 0 & 1 & 0 & \cdots \\
0 & 0 & 0 & 0 & \cdots \\
\vdots & \vdots & \vdots & \vdots & \cdots
\end{pmatrix}
$$

and

$$
G := F^t = \begin{pmatrix}
0 & 1 & 0 & 0 & \cdots & & \\
0 & 0 & 1 & 0 & \cdots & & \\
0 & 0 & 0 & 0 & 0 & 1 & 0 \cdots \\
\vdots & \vdots & \vdots & \vdots & \vdots & & \cdots
\end{pmatrix} .
$$

Again, we consider the element

$$
x := 1 + f \in End(V).
$$

One possible left inverse of x is given by z with the matrix representation

$$Z = \begin{pmatrix} 0 & 1 & -1 & 0 & 0 & 1 & 0 & \dots & -1 & \dots \\ 0 & 0 & 1 & 0 & 0 & -1 \dots & & & & \\ 0 & \dots & & & & & & & & \end{pmatrix}.$$

As before, the existence of the left-inverse z of x in $End(V)$ implies that every quasi-inverse of x is a left inverse of x. Again, we have to show that an arbitrary left inverse of x is not row-finite. We conduct the same case distinction as before.

Case 1 $z_{00} \neq 0$. Again, the condition $z_{00} + z_{01} = 1$ implies $z_{01} = 1 - z_{00}$. As before, we distinguish the cases $z_{00} = 1$ and $z_{00} \neq 1$.

Case 1.1 $z_{00} = 1$. This implies $z_{01} = 0$. We examine the second row and distinguish the cases $z_{10} = 0$ and $z_{10} \neq 0$.

Assume that $z_{10} = 0$. Because the product of the second row of Z with the first column of X has to vanish, this implies $z_{11} = 0$. Considering the product of the second row of Z with the second column of X, we get that $z_{12} = 1$. But then, considering the product of the second row of Z with the third column of X, we get that $z_{15} = -1$. Continuing, we see that the second row of Z consists of alternating entries of 1 and -1 at different positions ($i^2 + 1$ for $i \geq 1$).

Assume that $z_{10} \neq 0$. Because the product of the second row of Z and the first column of X has to vanish, this implies that $z_{11} = -z_{10}$. Now, we consider the product of the first row of Z and the second column of X.

This has to equal 1, hence $z_{12} = 2\lambda$. Considering further products of the second row of Z and the columns of X, we see that the second row of Z has infinitely many non-vanishing entries $\lambda, -\lambda, 2\lambda, -2\lambda, \ldots$.

Case 1.2: $z_{00} \neq 1, z_{00} \neq 0$. We need $z_{00} + z_{01} = 1$, so $z_{01} = 1 - z_{00} \neq 0$. Considering the product of the first row of Z with the other columns of X, we see that the first row of Z has infinitely many non-vanishing entries $\lambda, -\lambda, 2\lambda, -2\lambda, \ldots$; again, these non-vanishing entries are located at different locations in comparison with the situation of the usual shift operator.

Case 2: $z_{00} = 0$. It follows that $z_{01} = 1$. Therefore, the first row of Z has to have infinitely many non-vanishing alternating entries 1 and -1.

In every case, the matrix representation of Z with respect to Λ is not row-finite; consequently, there exists no operator in $End(V)$ which is adjoint to z. ◄

Conjecture 2.4.7. *The same statements should hold for an arbitrary skew field; but then, the case distinction of possible entries requires more work: A sequence of the form* $\lambda, -\lambda, 2\lambda, -2\lambda$ *will terminate.*

3 On Representability of Regular Involutive Rings

[...] and climbed up it like a mountain.
I established a base camp
about a third of the way across,[...].
It was a long and arduous journey,
full of trials, setbacks
and heroic derring-do on my part.

— From ONLY FORWARD
by Michael Marshall Smith

In this chapter, we complete the characterisation of representability of regular involutive rings. In [Nie03], $*$-Hermitian representations have been considered. In this chapter, we encompass symplectic representations and representations in vector spaces over skew fields of arbitrary characteristic.

We begin the chapter with a revision of a result given by Herstein in [Hst76]. Sadly, his presented proof was flawed. In this chapter, we will introduce an approach of Rowen [Row88] that leads to the desired results.

3.1 Revision of Herstein

In the beginning of his work [Hst76], Herstein tries to characterise faithful linear representability of primitive involutive rings. In particular, in [Hst76, Theorem 1.2.2], he claims the following

Claim 3.1.1. Let R be a primitive involutive ring with a minimal left ideal.

Then R has a linear representation $\sigma := (D, V_D, \langle \cdot, \cdot \rangle, \rho)$, where the form $\langle \cdot, \cdot \rangle$ is either $*$-Hermitian or symplectic.

A closer inspection of his proof shows that in fact, he has proven only the following.

Proposition 3.1.1. *Let R be a primitive involutive ring with minimal left ideal A.*

If R has a minimal projection, then R has a faithful linear $*$-Hermitian representation.

If R has no minimal projections, then the following holds: If I is an arbitrary minimal left ideal in R and $x \in I$, then we we can conclude that $xx^* = 0$ or $x^*x = 0$.

From his line of arguments, we can extract the following result:

Proposition 3.1.2. *Let R be a primitive involutive ring. Assume that R contains no minimal projections, but a minimal left ideal I with the property*

$$\forall x \in I. \qquad x^*x = 0.$$

Then R has a faithful linear symplectic representation.

3.2 Approach of Rowen

In his treatise **Ring Theory**, Rowen proves the following result.

Theorem 3.2.1. *[Row88, Theorem 2.13.21] Suppose R is a semiprime involutive ring with $Soc(R) \neq 0$.*

Then there exists a minimal non-vanishing left ideal L such that the following holds: Regarding L a right vector space V_D over the skew field $D := End_R(L)$, exactly one of the following conditions is satisfied.

1. *R has a linear $*$-Hermitian representation over V_D which is not symplectic.*

2. *R has a linear symplectic representation over D and D is a field.*

3. *$L^*L = 0$.*

Remark 3.2.1. The exact formulation in [Row88] is slightly different. For example, his use of the term *alternating* coincides with our term *symplectic* (see [Row88, Definition 2.13.20]). Furthermore, he speaks of L *having a (∗)-compatible form* $\langle \cdot, \cdot \rangle$, which means that $\langle \cdot, \cdot \rangle$ is non-degenerate and the adjoint with respect to $\langle \cdot, \cdot \rangle$ corresponds to the involution on R, i.e., L equipped with $\langle \cdot, \cdot \rangle$ is a representation space of the involutive ring R in our sense.[1]

In particular, on page 302, Rowen points out that his notation and approach include the case of characteristic two.

---

[1] The action of R on L is as usual given by left multiplication.

We need the following.

Lemma 3.2.2. *Let R be a primitive involutive ring and $e \neq 0$ a minimal non-Hermitian idempotent element in R.*
Then there exists $b \in R$ such that $e^* b e \neq 0$.

▷ Proof. The minimal left ideal Re generated by the idempotent e is a faithful left R-module. Since $e \neq 0$, we have $e^* \neq 0$. Hence, the action of e^* on Re cannot be trivial, i.e., there exists an element b such that $e^*(be) \neq 0$. ◁

Now we combine Section 1.3, Lemma 3.2.2 and Theorem 3.2.1.

Theorem 3.2.3. *Let R be a primitive regular involutive ring containing a minimal left ideal.*
Then R has a faithful linear $*$-Hermitian or symplectic representation.

▶ **Proof.** Since R is regular, R is semiprime by Lemma 1.3.10. Hence, the only thing left to show is that possibility (3) in Theorem 3.2.1 is excluded. But this is the case by Lemma 3.2.2. ◀

From Minimal Ideals to Representations

We can combine the previous results to get the following.

Corollary 3.2.4. Minimal ideal: $*$-Hermitian case
Let R be a regular involutive ring. Assume that there exists a primitive involutive regular extension $\widetilde{R}$ of R containing a minimal left ideal generated by a projection e.

Then R has a faithful linear $*$-Hermitian representation $\sigma = (V_D, D, \phi, \iota)$ over the skew field $D = e\tilde{R}e$.

Corollary 3.2.5. Minimal ideal: Symplectic case

Let R be a regular involutive ring. Assume that there exists a primitive involutive regular extension $\tilde{R}$ of R containing a minimal left ideal generated by an idempotent e but containing no minimal projections.

Then R has a faithful linear symplectic representation $\sigma = (V_D, D, \phi, \iota)$ over the skew field $D = e\tilde{R}e$.

From Representations to Minimal Ideals

In contrast to the previous section, we have the following.

Proposition 3.2.6. Construction of extension and ideal: Countable Hermitian case I

Let R be an at most countable regular involutive ring. Assume that $\sigma = (D, V_D, \phi, \rho)$ is a faithful linear $*$-Hermitian representation of R with $char(D) = 2$.

Assume furthermore that ϕ is not symplectic, i.e., there exists at least one anisotropic vector $v \in V_D$ and we are not in the pathological case.

Then there exists a primitive regular involutive extension $\tilde{R}$ of R containing a minimal left ideal A such that

1. A is generated by a projection e, i.e. $A = \tilde{R}e$,

2. $e\tilde{R}e$ is a skew field isomorphic to D.

Furthermore, the extension $\tilde{R}$ (and therefore R) admits a faithful linear $*$-Hermitian representation in a subspace W_D of V_D of at most countable dimension over the skew field D.

▶ **Proof.** We follow the line of arguments presented in [Nie03].

We fix the anisotropic vector $v \in V$. Starting with this vector, we construct a subspace W of V containing v of at most countable dimension such that R acts faithfully on W_D.

Then we make use of the generalised Gram-Schmidt-orthogonalisation (see [Gross79, Chapter 2]) to construct a decomposition of W into pairwise orthogonal non-degenerated subspaces of dimension one and two, i.e. into non-degenerated lines and planes

$$W_D = vD \oplus \bigoplus_{i \in I} w_i D \oplus \bigoplus_{j \in J} E_j$$

such that the fixed anisotropic vector v is the first basis vector of this decomposition.

Let Λ be the compatible basis with respect to this decomposition. Define the extension $\widetilde{R}$ by extending the ring $\iota(R)$ by all finite matrices with respect to Λ. Then, of course, the orthogonal projection p_v onto the one-dimensional subspace $U := span(v)$ is an element of $\widetilde{R}$. The projection $e := p_v$ is a generator of the minimal left ideal $A := \widetilde{R}e$ with the desired properties. ◀

Combining Proposition 3.2.6 and results given in [Nie03], we arrive at the following.

Proposition 3.2.7. Construction of extension and ideal: Countable Hermitian case II
Let R be an at most countable regular involutive ring with the faithful linear $*$-Hermitian representation $\sigma = (D, V_D, \phi, \rho)$ over a skew field of arbitrary characteristic.

Then there exists a primitive involutive regular extension $\widetilde{R}$ of R containing a minimal left ideal A such that

1. *A is generated by a projection e, i.e. $A = \widetilde{R}e$,*

2. *$e\widetilde{R}e$ is isomorphic to D.*

Furthermore, the extension $\widetilde{R}$ (and therefore R) admits a faithful linear $*$-Hermitian representation in a space W_D of at most countable dimension over the same skew field D.

Using ultraproducts, we can extend Proposition 3.2.7 to uncountable rings.

Proposition 3.2.8. Construction of extension and ideal: Uncountable Hermitian case

Let R be a non-countable regular involutive ring with the faithful linear $*$-Hermitian representation $\sigma = (D, V_D, \phi, \rho)$ over a skew field of arbitrary characteristic.

Then there exists a primitive involutive regular extension $\widetilde{R}$ of R containing a minimal left ideal A such that

1. *A is generated by a projection e, i.e. $A = \widetilde{R}e$,*

2. *$e\widetilde{R}e$ is isomorphic to D.*

Furthermore, the extension $\widetilde{R}$ (and therefore R) admits a faithful linear $*$-Hermitian representation in a space $W_{\widetilde{D}}$ over some ultrapower $\widetilde{D}$ of the skew field D.

▶ **Proof.** We can embed R into an ultraproduct of its countable regular involutive subrings R_i. Each R_i is, as a subring of R, representable in the representation space V_D over the skew field D. Since each R_i is countable, to each R_i there exists a primitive involutive regular extension $\widetilde{R}_i$ containing a minimal left ideal A_i such that A_i is generated by a projection e_i and $e_i \widetilde{R} e_i \cong D$.

Since the property of a regular ring R to have a minimal left ideal with these desired properties can be expressed in first order logic, the ultraproduct $\prod \widetilde{R}_i$ of the extensions $\widetilde{R}_i$ has this property, too. Hence, it admits a linear $*$-Hermitian representation. But we are cannot guarantee that the underlying skew field of the representation space is again D; all that can be said is that it is an ultraproduct of the skew fields of the representation spaces of the $\widetilde{R}_i$, i.e., an ultrapower of D.

Finally, the ultraproduct $\prod \widetilde{R}_i$ is an extension of $\prod R_i$ and therefore of R. ◄

We have similar results for symplectic representations.

Proposition 3.2.9. Construction of extension and ideal: Countable symplectic case

Let R be an at most countable regular involutive ring with the faithful linear symplectic representation $\sigma = (D, V_D, \phi, \rho)$.

Then there exists a primitive involutive regular extension $\widetilde{R}$ of R containing a minimal left ideal A such that

1. $A = \widetilde{R}e$ for some minimal idempotent e,

2. $e\widetilde{R}e$ is isomorphic to D,

3. $\forall x \in A. \qquad x^* x = 0$.

Furthermore, the extension $\widetilde{R}$ (and therefore R) admits a faithful linear symplectic representation in a space W_D of at most countable dimension over the same field D.

▶ **Proof.** Since R is at most countable, we can assume that V_D has at most countable dimension. Hence, we can assume that V_D has a decomposition $V_D = \bigoplus E_i$ into pairwise orthogonal non-degenerated planes.

Let Λ be the compatible basis with respect to this decomposition. Define the extension $\widetilde{R}$ in the usual way, i.e., $\widetilde{R} := R + J$, where J denotes the set of all finite matrices with respect to Λ. Then $\widetilde{R}$ is a primitive regular involutive extension of R and $\widetilde{R}$ is representable over V_D.

Now, let v be the basis vector of the plane E_0, i.e. the first element of the compatible basis Λ. Let $e : V_D \to V_D$ be the idempotent map onto $span(v)$, that is, on the elements of Λ, the morphism e acts in the following way.

$$e : \Lambda \to V \qquad e_i \mapsto \begin{cases} e_i & \text{if } i = 0 \ (e_i = v) \\ 0 & \text{else } (e_i \neq v) \end{cases}$$

In particular, the matrix representation of e with respect to Λ has the form

$$[e]^\Lambda = diag(1, 0, 0, \ldots,)$$

Note that the map e is not a projection in the sense that $e = e^* = e^2$: Since the form ϕ is symplectic, e is not Hermitian. The adjoint e^* of e has the matrix representation

$$[e^*]^\Lambda = diag(0, 1, 0, 0, \ldots,)$$

Now, consider the minimal left ideal $A := \widetilde{R}e$ generated by the idempotent e.

A consists of all matrices of the form

$$
a = \begin{pmatrix}
a_{11} & 0 & 0 & \cdots \\
a_{21} & 0 & 0 & \cdots \\
a_{31} & 0 & 0 & \cdots \\
\vdots & \cdots & \cdots & \cdots
\end{pmatrix}
$$

with $a_{ni} \in D$, but of course only finitely many entries different from zero.

Then, using the decomposition of V_D into a orthogonal sum of non-degenerated planes, we can calculate the adjoint of an element $a \in A$ and get

$$
a^* = \begin{pmatrix}
0 & 0 & 0 & 0 & \cdots \\
-a_{21} & a_{11} & -a_{41} & +a_{31} & \cdots \\
0 & 0 & 0 & 0 & \cdots \\
\vdots & \cdots & \cdots & \cdots & \cdots
\end{pmatrix}
$$

Therefore, for an arbitrary element $a \in A = Re$, we get $a^*a = 0$, independent of the characteristic of the field D. ◄

Proposition 3.2.10. Construction of extension and ideal: Uncountable symplectic case

Let R be a non-countable regular involutive ring with the faithful linear symplectic representation $\sigma = (D, V_D, \phi, \rho)$.

Then there exists a primitive involutive extension $\widetilde{R}$ of R containing a minimal left ideal A such that

1. *$A = \widetilde{R}e$ for some minimal idempotent e,*

2. *$e\widetilde{R}e$ is isomorphic to D,*

3. $\forall x \in A. \qquad x^* x = 0.$

Furthermore, the extension $\widetilde{R}$ (and therefore R) admits a faithful linear symplectic representation in a space $W_{\widetilde{D}}$ over some ultrapower $\widetilde{D}$ of the skew field D.

▶ **Proof.** We can embed R into an ultraproduct of its countable regular involutive subrings R_i. Each R_i is, as a subring of R, representable over the representation space V_D. Since each R_i is countable, to each R_i there exists a primitive involutive regular extension $\widetilde{R}_i$ containing a minimal left ideal A_i generated by an idempotent e_i such that $e_i \widetilde{R} e_i \cong D$.

Since the property of a regular involutive ring to have a minimal left ideal with the desired properties can be expressed in first order logic, the ultraproduct $\prod \widetilde{R}_i$ of the extensions $\widetilde{R}_i$ has a minimal left ideal with the stated properties, too. Hence, it admits a linear symplectic representation. Again, the underlying skew field of the representation space is an ultrapower of D.

Finally, the ultraproduct $\prod \widetilde{R}_i$ is an extension of $\prod R_i$ and therefore of R. ◄

3.3 Approximation

In this section, we want to relate faithful linear representations of involutive rings with a certain process of approximation.

Concerning $*$-regular rings that admit a faithful linear positive $*$-Hermitian representation, we have the following result.

Theorem 3.3.1. *[Mic03, Theorem 3.15] Let R be a $*$-regular ring with the faithful linear positive representation $\sigma = (D, V_D, \phi, \rho)$ in a vector space V of countable dimension over D.*

Then R is a homomorphic image of a subring of a product of $*$-regular Artinian rings.

Our aim is to generalise this result to regular involutive rings that admit faithful linear $*$-Hermitian and faithful linear symplectic representations. We even want to generalise the result to representations in spaces of arbitrary dimension. This is, we want to show that an involutive ring R with a linear representation, either $*$-Hermitian or symplectic, is a homomorphic image of a subring of a product of Artinian involutive rings.

To reach this result, we will focus on representations in a vector space of at most countable dimension first. Then we use ultraproducts to generalise the result to representations in spaces of arbitrary dimensions.

Representations of at most Countable Dimension

For this section, we assume that R has a faithful linear $*$-Hermitian or symplectic representation $\sigma = (D, V_D, \phi, \rho)$, where V_D has at most countable dimension over D.

Basis and Projections Since V_D is of at most countable dimension and ϕ is a non-degenerated ($*$-Hermitian or symplectic) form, we can decompose V_D into a direct sum

$$V_D = \bigoplus_{i \in I} U_i \oplus \bigoplus_{j \in J} F_j,$$

where each summand U_i is a anisotropic line and each summand F_j a metabolic plane. In particular, each summand is non-degenerated.

Remark 3.3.1. In the case of a symplectic representation $I = \emptyset$, while in the case of a *-Hermitian non-symplectic representation $I \neq \emptyset$.

Since V_D is of at most countable dimension, we have at most countably many infinite summands. If we drop the distinction between the two possible types of summands, we can write $V_D = \bigoplus_{i \in I} W_i$, where I is a finite or infinite subset of $\mathbb{N}$ and W_i a (non-degenerated) subspace of dimension one or two. We take β to be the compatible basis of V_D corresponding to this decomposition.

We define an ordered family of projections in the following way: For each $n \in I$, we set p_n to be the projection on the subspace consisting of the first n summands:

$$p_n : V_D \to V_D \qquad v \mapsto \begin{cases} v & \text{if } v \in \bigoplus_{i=0}^{n} U_i \\ 0 & \text{else} \end{cases}$$

Then each projection is indeed a projection in the sense that

$$p_n = p_n^2 = p_n^*$$

Remark 3.3.2. In general, the maps p_n are *not* projections on n-dimensional subspaces: The dimension of the range of p_n is the subspace of the first n summands in the given decomposition. But we have to proceed in this way to guarantee that the maps are Hermitian: If the first summand W_0 would be a metabolic plane and p_0 the idempotent map onto the first basis vector of W_0, then p_0 would not be Hermitian. But the idempotent map onto W_0 would indeed be Hermitian, i.e. a projection with two-dimensional range.

Artinian Factors   We define the set

$$M_n(D) := \{a \in End(V_D) : p_n a p_n = a\}$$

consisting of the endomorphisms on V_D whose matrix representation with respect to β has a square block in the upper left corner and zeros elsewhere. We note the following.

1. For every $n \in \mathbb{N}$, the set $M_n(D)$ is an involutive ring on its own.

2. The matrix representation of $a \in End(V_D)$ with respect to β is a countable small matrix iff the following holds.

$$\forall k \in \mathbb{N}.\exists l \in \mathbb{N}. \qquad (p_l a p_k = a p_k) \wedge (p_k a p_l = p_k a)$$

Remark 3.3.3. Even though this setting is formally similar to the one in [Mic03], the situation is a bit different: Each endomorphism that belongs to $M_n(D)$ has a matrix representation with a square block in the upper left corner – albeit the size of this block is not necessarily $n \times n$, but depends on the decomposition of the vector space.

Product We define the product

$$S := \prod_{n \in \mathbb{N}} M_n(D) = \{(a_n)_{n \in \mathbb{N}} : a_n \in M_n(D)\}.$$

Of course, the product ring S is an involutive ring.

For an element $(a_n) \in S$ and an element $a \in End(V_D)$, we say that (a_n) *converges to* a if the following holds:

$$\forall l \in \mathbb{N}.\exists n \in \mathbb{N}.\forall m \geq n. \qquad a_m p_l = a p_l \ \& \ p_l a_m = p_l a$$

Lemma 3.3.2. *In this situation, the following hold.*

1. *If $(a_n) \in S$ is a convergent series, the limit is unique.*

2. *If $(a_n) \in S$ is a convergent series, the limit is a countable small matrix, hence it has an adjoint.*

3. *For any endomorphism $a \in End(V_D)$ whose matrix representation with respect to α is a countable small matrix, there is a series $(a_n) \in S$ which converges to a.*

▷ Proof. This result corresponds to [Mic03, Proposition 3.14]. In the presented proof, she makes use of the projections p_n only. ◁

Subspace We define the subset U of S to be the set of all convergent sequences whose limit lies in the image of R, i.e.,

$$U := \{(a_n) \in S : (a_n) \text{ converges to some } a \in \rho(R)\}.$$

Lemma 3.3.3. *In this situation, the following hold.*

1. *U forms an involutive subring of S.*

2. *The map $\varphi : U \to \rho(R)$ mapping a sequence to its limit is a surjective $*$-ring homomorphism.*

▷ Proof.

1. The involution on U is inherited from the involution on S. All ring operations (addition, multiplication and involution) are compatible with the limit process.

2. For $a \in \rho(R)$, we set $a_n := p_n a p_n$ to get a sequence (a_n) in S. This sequence converges to a in $\rho(R)$, so (a_n) lies in U. Since taking the limit is compatible with all operations, the map φ is a surjective $*$-ring homomorphism.

◁

Conclusion By the above construction, we have connected the property of a ring to have a faithful linear $*$-Hermitian or symplectic representation in vector space of at most countable dimension with the property to be a homomorphic image of a subring of a product of Artinian involutive rings, i.e., we have shown the following result.

Proposition 3.3.4. *Let R be an involutive ring with a faithful linear $*$-Hermitian or symplectic representation in a vector space of at most countable dimension.*

Then R is a homomorphic image of a subring of a product of Artinian involutive rings.

Representations of Arbitrary Dimension

We begin with a corollary for countable rings.

Corollary 3.3.5. *Let R be a countable involutive ring having a faithful linear $*$-Hermitian or symplectic representation.*

Then R is a homomorphic image of a subring of a product of involutive Artinian rings.

▷ Proof. Since R is countable, we can construct a representation in a vector space of at most countable dimension. Then we can apply Proposition 3.3.4. ◁

Now, we can approach the desired result.

Theorem 3.3.6. *Let R be an involutive ring having a faithful linear $*$-Hermitian or symplectic representation $\sigma = (D, V_D, \phi, \rho)$.*

Then R is a homomorphic image of a subring of a product of involutive Artinian rings.

▶ **Proof.** If R is countable, we are done by Corollary 3.3.5.

So, assume R is not countable. Following [Nie03], we can embed R into an ultraproduct of countable involutive subrings R_i of R. As a subring of R, each R_i admits a faithful linear ($*$-Hermitian or symplectic[2]) representation in V_D.

Consider a fixed index $i \in I$. Since R_i is countable, Corollary 3.3.5 yields the existence of an involutive sub-

---

[2]Where of course, the type of the representation is fixed by the type of the representation of R.

ring U_i a product ring $\prod M_n^i(D)$ such that R_i is a homomorphic image of U_i. Since the index $i \in I$ was arbitrary, this holds for each $i \in I$. Taking the product of the subrings U_i, we get a surjective ring homomorphism

$$f : \prod_{i \in I} U_i \to \prod_{i \in I} R_i.$$

That is, $\prod R_i$ is a homomorphic image of a subring of $\prod_{i \in I} \left(\prod_{n \in \mathbb{N}} M_n^i(D) \right)$.

Let θ be an ultrafilter on I such that R is isomorphic to a subring of $\prod R_i / \theta$. We identify R with its isomorphic copy in the ultraproduct. We get a composition of maps

$$\prod_{i \in I} U_i \xrightarrow{f} \prod_{i \in I} R_i \xrightarrow{\pi_\theta} \prod_{i \in I} R_i / \theta.$$

Taking preimages of elements of R under the map

$$\pi_\theta \circ f : \prod_{i \in I} U_i \to \prod_{i \in I} R_i / \theta,$$

we get an involutive subring $U' \leq \prod U_i$ such that the map $f' : U' \to R$ is a surjective ring-homomorphism. Therefore, R is a homomorphic image of an involutive subring of $\prod_{i \in I} \left(\prod_{n \in \mathbb{N}} M_n^i(D) \right)$. ◄

3.4 Atomic Extensions

In this section, we will show that every (regular) involutive ring that admits a faithful linear representation has an atomic (regular) involutive extension that also admits a faithful linear representation. We will first consider the case of at most countable rings before we generalise the result to non-countable rings. For the first case, we will use the construction of a representation in a subspace of at most countable dimension. For the second, we have to work with ultraproducts of countable substructures.

Spaces of Countable Dimension

Proposition 3.4.1. *Let R be a regular involutive ring with the faithful linear representation $\sigma = (D, V_D, \phi, \rho)$, where the form ϕ is either $*$-Hermitian or symplectic and the vector space V_D has at most countable dimension over D.*

Then R has an atomic regular involutive extension $\widetilde{R}$ which admits a faithful linear representation in the same vector space V_D. In particular, the representation of $\widetilde{R}$ is of the same type as σ.

▶ **Proof.** Since V_D has at most countable dimension, we can construct a suprabolic decomposition of V_D and introduce the compatible basis Λ with respect to that decomposition.

We define the extension $\widetilde{R}$ as usual: We take the ring generated by $\rho(R)$ and the set of all finite matrices with respect to Λ. Then $\widetilde{R}$ is the desired extension:

1. Since it contains all finite matrices, it is primitive and atomic.

2. Since it is generated by R and all finite matrices with respect to the compatible basis Λ, it is regular.

3. It admits a faithful linear representation in the space V_D.

◀

Countable Rings

Corollary 3.4.2. *Let R be a countable regular involutive ring with a faithful linear representation, either $*$-Hermitian or symplectic.*

Then R has an atomic regular involutive extension $\widetilde{R}$ that admits a faithful linear representation in the same representation space.

▷ Proof. Since R is countable, we can assume that V_D has at most countable dimension. Hence, Proposition 3.4.1 applies. ◁

Uncountable Rings

Proposition 3.4.3. *Let R be a uncountable regular involutive ring with the faithful linear $*$-Hermitian or symplectic representation σ.*

Then R has an atomic regular involutive extension $\widetilde{R}$ which admits a faithful linear representation. The representation of the extension $\widetilde{R}$ is of the same type as σ.

▶ **Proof.** We can embed R into an ultraproduct $\prod R_i/\theta$ of its countable regular involutive subrings R_i, θ an ultrafilter on I. As a subring of R, each R_i admits a faithful linear representation, where the representation type is fixed by the type of the representation of R.

Fix an index $i \in I$. We can apply Corollary 3.4.2 to conclude that R_i has an atomic extension $\widetilde{R}_i$, where $\widetilde{R}_i$ has a faithful linear representation in the same space as R_i. Since $\widetilde{R}_i$ is an extension of R_i, the corresponding ultraproduct $\prod \widetilde{R}_i / \theta$ is an extension of $\prod R_i / \theta$, hence of R.

Now, we use that the property of a ring to have a faithful linear representation can be expressed in first-order logic. Furthermore, the property to be atomic can be expressed in first-order logic, too.

Since each $\widetilde{R}_i$ is an atomic involutive ring with a faithful linear representation, so is the ultraproduct $\prod \widetilde{R}_i / \theta$. Thus, we are done. ◄

4 On *-Regular Rings

"It's okay", Hiro says.
"I'm sure they'll listen to Reason."

— From SNOW CRASH
by Neal Stephenson

This chapter focuses on *-regular rings. The aim is to show that every *-regular ring is representable, that is, every *-regular ring is a subring of a product of endomorphism rings.

We start by introducing the necessary notation. In Section 4.2, we show that the class of all *-regular rings is a variety and that the property of a *-regular ring to admit a faithful linear positive representation is closed under substructures, ultraproducts and homomorphic images. In Sections 4.3 and 4.4, we develop the particular technique needed for this chapter. In Section 4.5, we prove that every subdirectly irreducible *-regular ring has a faithful linear positive representation. In Section 4.6, we show that every variety of *-regular rings is generated by its simple Artinian members.

4.1 Modules and Morphisms

The following notation and conventions will be used in this chapter only.

We consider right modules over rings. Modules will be denoted by M_S, N_T. Submodules will be denoted by M_i, N_i, without mentioning the underlying ring. If we do not explicitly state otherwise, we assume that the underlying ring is *-regular.

Morphisms between modules M_S and N_T and endomorphisms of M_S are denoted by non-indexed symbols φ, ψ. Morphisms between submodules $M_i, M_j \leq M$ are denoted by $\varphi_{ji} : M_i \to M_j$, where the indices should be read from right to left. If M_i, M_j have trivial intersection, we define the graph of a morphism $\varphi_{ji} : M_i \to M_j$ by $\Gamma(\varphi_{ji}) = \{x - \varphi_{ji}(x) : x \in M_i\}$.

Observation 4.1.1. Let $M_i \cap M_j = \{0\}$. Note that $\Gamma(\varphi_{ji})$ is a relative complement of M_j in $[0, M_i + M_j]$ and, conversely, each such relative complement gives rise to a morphism $\psi_{ji} : M_i \to M_j$.

For a partial morphism between modules M_S and N_T (between submodules M_i and N_j), we write $\varphi :: M \to N$ ($\psi_{ji} :: M_i \to M_j$), where a partial morphism is defined only on a submodule $dom(\varphi) \leq M$ ($dom(\psi_{ji}) \leq M_i$) and $\varphi : dom(\varphi) \to N$ ($\psi_{ji} : dom(\psi_{ji}) \to M_j$) is a module morphism.

Observation 4.1.2. Let M be a module with a direct decomposition $M = \bigoplus M_i$ and denote the corresponding projections and embeddings by π_i and ε_j, respectively.

Consider a morphism $\varphi_{ji} : M_i \to M_j$. Then the composition of φ_{ji} with the projection $\pi_i : M \to M_i$ yields a morphism $\varphi_{ji} \circ \pi_i : M \to M_j$ defined on all of M, i.e., $\varphi_{ji} \circ \pi_i$ is an element of $Hom(M, M_i)$. Since $M_i \leq M$, we can consider $\varphi_{ji} \circ \pi_i$ as an element of $End(M)$, too. For the latter point of view, the formally correct approach would be to consider $\varepsilon_j \circ \varphi_{ji} \circ \pi_i$. To avoid technical and notational overload, we will treat $\varphi_{ji} \circ \pi_i$ itself as an element of $End(M)$. Note that the composition $\varphi_{ji} \circ \pi_i$ is nothing else than the extension of the map $\varphi_{ji} : M_i \to M_j$ to all of M by defining the action of the extension to be trivial on the other summands of M. Very rarely, we write $\overline{\varphi_{ji}}$ for this extension: We just use overlined symbols if we want to distinguish between a partial map and its extension.

Conversely, consider a morphism $\varphi \in M$. We introduce maps φ_i and φ_{ji}, defined by compositions of φ with embeddings and projections.

$$\varphi_i := \varphi \circ \epsilon_i : M_i \to M \quad \varphi_{ji} := \pi_j \circ \varphi_i = \pi_j \circ \varphi \circ \epsilon_i : M_i \to M_j$$

We notice that each $\varphi \in End(M)$ can be decomposed in the following ways.

$$\varphi = \bigoplus_{i \in I} \varphi_i = \bigoplus_{i \in I} \sum_{j \in I} \varphi_{ji}$$

Hence, there is a 1-1-correspondence between

1. A morphism $\varphi : M \to M$,

2. a family $\{\varphi_i : i \in I\}$, where $\varphi_i : M_i \to M$, and

3. a family $\{\varphi_{ji} : i, j \in I\}$, where $\varphi_{ji} : M_i \to M_j$.

We agree to write $\varphi = \sum_{i,j \in I} \varphi_{ji}$, with the convention

stated above. We agree to not impose a rigorous notational strictness, but to understand the notation in the
natural sense. Similar to the observation above, we note
than $\varphi_i \circ \pi_i$ is nothing else that the extension of the map
$\varphi_i = \varphi \circ \epsilon_i : M_i \to M$ to whole M.

We note that the conventions are compatible with ad-
dition and multiplication: We can form the sum $\varphi_{ji} + \psi_{ji}$
and the composition $\varphi_{jk} \circ \psi_{ki}$ in the natural sense; for
$\varphi, \psi \in End(M)$, we have the following compatibilities.

$$(\varphi + \psi)_{ji} = \varphi_{ji} + \psi_{ji} \qquad (\varphi \circ \psi)_{ji} = \sum_{k \in I} \varphi_{jk} \circ \psi_{ki}$$

We agree to write $1 = id_M : M \to M$. Then we have
$1_{ii} = id_{M_i} : M_i \to M_i$ (that is, the corresponding extension of 1_{ii} acts like the identity on M_i and trivially
on every other summand M_j) and $1_{ji} = 0_{ji} : M_i \to M_j$
(that is, the extension of 1_{ji} coincides with the zero map
on M).

Observation 4.1.3. For cyclic modules M_S, N_S with
generators x, y, a morphism $\varphi : M \to N$ is determined
by its action on the generator x of M. If $M = xS$, we have
$f(xs) = f(x)s$ for every $xs \in M$. Conversely, each choice
$z \in yS$ defines a morphism $g : M \to N$ via $xs \mapsto zs$.

In particular, let R be a regular ring and consider the
module R_R. Assume that $I = eR, J = fR$ are principal right ideals in R (that is, cyclic submodules of R_R).
Since R is regular, the generators e, f can be taken to be
idempotent. Let $r \in R$ such that $re \in J$, that is, $re = fc$
for some $c \in R$ (or, equivalently, $f(re) = re$).

Then the left multiplication with r defines a right-R-
module-homomorphism $\hat{r}$ between I and J in the follow-

ing way.

$$\widehat{r} : I \to J \qquad es \mapsto r(es)$$

From now on, if possible, we denote the action defined by left multiplication with an element r by $\widehat{r}$. We will speak of the *left multiplication morphism* (or *left multiplication map* or *left multiplication*) $\widehat{r}$.

4.2 The Variety of ∗-Regular Rings

This section deals with the class $\mathcal{R}$ of all ∗-regular rings. In particular, we want to show that $\mathcal{R}$ is a variety, i.e., that it is closed under products, substructures and homomorphic images.

Subsequently, we introduce the notion of a directed union of ∗-regular rings and a result that will be helpful if we consider ∗-regular rings without unit.

We finish the section with the consideration of substructures, ultraproducts and homomorphic images of representable ∗-regular rings.

∗-Regular Rings and Class Operators

It is obvious that an arbitrary product of ∗-regular rings, with the operations defined componentwise, is again a ∗-regular ring.

For homomorphic images, we follow the line of arguments given in [Mic03].

Lemma 4.2.1. *A two-sided ideal of a ∗-regular ring is ∗-regular.*

▷ Proof. By [Good91, Lemma 1.3], a two-sided ideal of a

regular ring is itself regular. By [Mic03, Proposition 1.7], a two-sided ideal of a ∗-regular ring is closed under the involution of the ring. ◁

For substructures, we recall the notion of the *Rickart relative inverse* of an element of a ∗-regular ring. Some preliminary work is needed.

Definition 4.2.2. Left and right projection
Let R be a ∗-regular ring. For an element $a \in R$, we call the unique projection e in R that generates the principal right ideal aR the *left projection of* a and the unique projection f in R that generates the principal left ideal Ra the *right projection of* a.

Remark 4.2.1. This notation is due to the fact that a is left-invariant under e and right-invariant under f, that is, we have $ea = a = af$. We denote the left and right projection of a by $l(a)$ and $r(a)$, respectively.

If R has a unit, we have $ann_R^l(a) = R(1 - e)$ and $ann_R^r(a) = (1 - f)R$.

Remark 4.2.2. This terminology can be found in either [Kap68], p. 27–28 or [Kap55], p. 525.

In a ∗-regular ring, the left and right projection of an element can be constructed explicitly.

Lemma 4.2.3. *Let R be a ∗-regular ring, $x \in R$ and let x' denote any quasi-inverse of x in R.*

Then the left and right projection of x are given in the following way.

$$l(x) := x(x^*x)'x^* \qquad r(x) := x^*(xx^*)'x$$

▷ **Proof.** See [Mic03], p. 9–10. Note that there is a slight difference in notation: While Kaplansky calls e the left projection of x (since x is left-invariant under e), Micol calls e the right projection of x (since e generates the same principal right ideal). ◁

Lemma 4.2.4. *Let R be a ∗-regular ring.*

Then for each element $a \in R$ there exists a unique element $q(a)$ such that the following conditions hold.

1. $e := l(a) = aq(a^*a)a^*$

2. $f := r(a) = a^*q(aa^*)a$

3. $fq(a) = q(a)$

4. $aq(a) = e$

Furthermore, $q(a)$ has the properties that $q(a)a = f$, the left projection of $q(a)$ is f and the right projection of $q(a)$ is e.

▷ **Proof.** This result is due to Kaplansky (see [Kap68] or [Kap55]). Here, we have defined a function $q : R \to R$ that maps each $a \in R$ to the unique element y with the listed properties. ◁

Remark 4.2.3. We call $q(a)$ the *relative inverse of a*. We note that a is the relative inverse of $q(a)$, so $q^2 = id_R$.

With the above results about the relative inverse, we can prove the following.

Proposition 4.2.5. *Let R be a ∗-regular ring and $S \leq R$ be a ∗-subring of R.*

Then S is ∗-regular if and only if S is a closed under the operation $q : R \to R$.

▶ **Proof.** Assume that S is closed under q. For an element $a \in S$, the map q gives a quasi-inverse $q(a)$ of a, so S is regular. Since S is a *-subring of the *-regular subring R, S is itself *-regular.

Conversely, assume that $S \leq R$ is a *-regular subring of R. Let $x \in S$. Due to Lemma 4.2.3, we can construct the left and the right projection of x within S, using the involution on S and any quasi-inverses of x, x^*, xx^*, x^*x in S. By Lemma 4.2.4, there exists an element y with the desired properties within the *-regular ring S. Since y is the unique element with these properties, we have $y = q(x)$. Hence, S is closed under q. ◀

Consequently, we have arrived at the following desired result.

Theorem 4.2.6. *The class $\mathcal{R}$ of *-regular rings is closed under products, homomorphic images and q-substructures, i.e., it forms a variety.*

*Speaking about the variety of *-regular rings, we consider the map $q : R \to R$ as a unary operation on R and part of the structure, that is a *-regular ring R (without unit) will be considered as an algebra of type $(R, +, \cdot, ^*, q, 0)$.*

Remark 4.2.4. Notice that the left and the right projection of an element x of a *-regular ring R are given explicitly in terms involving only x, x^* and quasi-inverses of products of x and x^*. Hence, the relative inverse of x is preserved by a morphism $\phi : R \to S$ of *-regular rings.

Of course, for a product of *-regular rings R_i, the relative inverse is given componentwise.

Directed Unions and Rings without Unit

Definition 4.2.7. Directed union of rings

Let R be a ring and $\mathcal{S} = \{S_i : i \in I\}$ be a directed family of subrings of R. We say that R is a *directed union* of the family $\mathcal{S}$ if for each $r \in R$ there exists $k \in I$ such that $r \in S_k$.

Remark 4.2.5. Casually, we speak of R *being the directed union of the* S_i, without giving the family of the S_i an extra name, and we write $R = \bigcup_{i \in I} S_i$, using the usual symbol for an ordinary union. Of course, an arbitrary union of rings is in general not a ring; hence, the lax notion does not lead to the risk of misunderstandings.

Lemma 4.2.8. *Let R be a ∗-regular ring and assume that R is the directed union of a family $\mathcal{S}$ of ∗-regular subrings S_i of R.*

Then R is a ∗-regular subring of an ultraproduct of the rings S_i, $i \in I$.

▷ Proof. Since the class of all ∗-regular rings forms a variety, this follows from [Gor98], Theorem 1.2.12 (1). ◁

Representability and Universal Algebra

Lemma 4.2.9. *Let R be a ∗-regular ring with a representation $\sigma = (D, V_D, \langle \cdot, \cdot \rangle, \rho)$.*

Then every ∗-regular subring S of R is representable. If the representation of R is faithful, so is the representation of S.

▷ Proof. Just take the restriction $\rho_{|_S}$. ◁

Proposition 4.2.10. *Let $\{S_i : i \in I\}$ be a family of linearly representable *-regular rings over an arbitrary index set I. Let U be an ultrafilter on I.*

Then the ultraproduct

$$R := (\prod_{i \in I} S_i)/U$$

is linearly representable. If every S_i has a faithful linear representation, then so does R.

▶ **Proof.** Consider the class of 2-sorted structures

$$\mathcal{K}: \quad = \quad \{(R, V) : R \text{ a *-regular ring, } V \text{ a vector space}$$
$$\text{s.t.} \quad R \text{ has a linear positive representation in } V\}$$

Since the relation that the *-regular ring R has a (faithful) linear positive representation in the vector V can be expressed in first-order logic, an ultraproduct of a family of structures $(R_i, V_i) \in \mathcal{K}$ lies again in $\mathcal{K}$. ◀

Proposition 4.2.11. *Let R and S be *-regular rings. Assume that R has a faithful linear representation and S is a homomorphic image of R.*

Then S has a faithful linear representation.

▶ **Proof.** In [Mic03, Theorem 3.8], Micol has shown that any homomorphic image of a g-representable *-regular ring with unit is g-representable. We adapt the given proof to the situation of *-regular rings without unit, but focus our attention on the case that R has a linear representation.

Assume that $h := \rho$ is the *-ring embedding given by the faithful representation of R and let J a two-sided ideal in R. The map h can be factorised by the quotient map π_J associated with J if and only if $J \subseteq ker(h)$. Hence, if $J \neq \{0\}$, the embedding h cannot be factorised by π_J.

The idea is to find a $*$-morphism $g : R \rightarrow End(V_D, \langle .,. \rangle)$ such that $J = ker(g)$. Then g factorised by π_J is a $*$-ring-embedding. As in [Mic03], the existence of such a morphism g can be shown if R is faithfully representable. We define a multi-sorted structure $\mathcal{M}_{R,J}$ as follows.

Sets. We have two sets: A set V of vectors and a set D of scalars.

Operations and relations. The operations and relations consist of group operations for V, ring operations for D, a map $* : D \rightarrow D$, a map $. : V \times D \rightarrow V$, a map $\langle \cdot, \cdot \rangle : V \times V \rightarrow D$, for each $r \in R$, a map $\rho(r) : V \rightarrow V$, and a subset $P \subseteq V$.

Axioms. We require that $\mathcal{M}_{R,J}$ satisfies the following axioms: V is an Abelian group, $(D,^*)$ is an involutive skew field, $.$ is unital, $\langle \cdot, \cdot \rangle$ is a scalar product and the map $\rho : R \rightarrow (End(V_D), \langle \cdot, \cdot \rangle)$ defined by $\rho : r \mapsto \rho(r)$ is a faithful linear positive representation of R.

These axioms contain the following formulae.

$$\forall d \in D. \quad d \neq 0 \Rightarrow \exists e \in D. de = 1 \qquad (4.1)$$

$$\forall x, y \in V. \exists d \in D. \qquad \langle x, y \rangle = d \qquad (4.2)$$

$$\langle \cdot, \cdot \rangle : V \times V \rightarrow D \text{ is a scalar product} \qquad (4.3)$$

For all $r, s \in R$ with $r \neq s$:

$$\exists v \in V. \qquad \rho(r)v \neq \rho(s)v \qquad (4.4)$$

For $r \in R$:

$$\forall x, y \in V. \qquad \langle \rho(r)x, y \rangle = \langle x, \rho(r^*)y \rangle \qquad (4.5)$$

Now, consider the following formulae.

For $f \in J$:

$$\forall x \in V. \qquad (x \in P \Rightarrow \rho(f)(x) = 0) \qquad (4.6)$$

For $r \in R \setminus J$:

$$\exists x \in V. \quad x \in P \wedge \rho(r)(x) \neq 0 \qquad (4.7)$$

Lemma 4.2.12. *Let R be a *-regular ring and J a two-sided ideal of R.*

If R is faithfully representable, then there exists a structure $\mathcal{M}_{R,J}$, satisfying the above axioms and formulae.

▷ **Proof.** Let $\sigma' = (D', V'_{D'}, \langle \cdot, \cdot \rangle', \rho')$ be the given representation of R. As in [Mic03], Lemma 3.10, we make the natural identifications, i.e., $D := D', V := V'$ etc. Hence, we have a structure with the desired sorts, axioms and operations. Obviously, Conditions 4.1, 4.2, 4.3, 4.4, 4.5 are satisfied. Hence, all formulae, possibly except that of type 4.6 and 4.7, are satisfied, for arbitrary $P \subseteq V$.

Now we consider a finite set $\mathcal{T}$ of formulae of type 4.6 and 4.7. We call F the (finite) set of $f \in J$ involved in

T. We set

$$P := \{x \in V : \forall f \in F.\ \rho(f)(x) = 0\}.$$

We have to show that the formulae of T hold, that is, for an element $r \in R \setminus J$ involved in T, we have to find $x \in V$ such that $x \in P$ and $\rho(r)(x) \neq 0$.

We consider the left ideal $\langle F \rangle_{l,R}$ generated by F in R. Since R is ∗-regular, $\langle F \rangle_{l,R}$ is generated by a projection $e \in R$. Since $F \subseteq J$, we have $e \in J$. Furthermore, for all $f \in F$, the following hold.

1. $Rf \subseteq Re$.

2. $fe = f$, since $e^2 = e$.

3. $\rho(f) \circ \rho(e) = \rho(f)$, since ρ is a morphism of rings.

Now, for $r \in R \setminus J$, there exists a projection $p \in R$ with $rp = r$. If we consider the two projections p and e in R, there exists a supremum of p and e in the projection lattice of R. Let us denote this supremum by $a := p \vee e$.

Then we have $pa = p$ and $ea = e$, so in particular $ra = (rp)a = r(pa) = rp = r$. Thus, we can write r in the following way.

$$r = ra = ra - re + re = r(a - e) + re$$

We notice that $r(a-e) \neq 0$: If we would have $r(a-e) = 0$, we could conclude that $r = re \in J$, a contradiction. Hence, since ρ is injective, we can conclude that there exists $m \in V$ such that

$$\big(\rho(r)(\rho(a) - \rho(e))\big)(m) \neq 0.$$

We set
$$x := \big(\rho(a) - \rho(e)\big)(m).$$

Then of course, $\rho(r)(x) \neq 0$. Furthermore, for arbitrary $f \in F$, we have

$$
\begin{aligned}
\rho(f)(x) &= \big(\rho(f) \circ \rho(e)\big)(x) \\
&= \big(\rho(f) \circ \rho(e)\big)\big((\rho(a) - \rho(e))(m)\big) \\
&= \big(\rho(f) \circ \rho(e) \circ (\rho(a) - \rho(e))\big)(m) \\
&= \big(\rho(f) \circ (\rho(ea) - \rho(e))\big)(m) \\
&= \big(\rho(f) \circ (\rho(e) - \rho(e))(m) = 0.
\end{aligned}
$$

Hence, $x \in P$. By the compactness theorem, we can conclude that there exists a desired structure $\mathcal{M}_{R,J}$ (see [Mic03], Lemma 3.10). ◁

Lemma 4.2.13. *If $\mathcal{M}_{R,J}$ exists, then the factor ring R/J is faithfully representable.*

▷ Proof. The proof of [Mic03], Lemma 3.11 does not make any use of the existence of a neutral element in R; therefore, it applies to our situation as well. ◁ ◀

4.3 General Framework

In this section, we will develop the necessary machinery for the proof of the main result. In order to simplify the line of arguments and to clarify the applied technique, we have chosen to separate the ring-theoretical aspects, the lattice- and frame-theoretical aspects and the general module-theoretical mechanisms as far as possible. The presented approach allows us to emphasise which conditions are used in each situation. For example, the notion of an abstract matrix ring can be based entirely on a decomposition system and a fixed subring $C \leq End(M_0)$, without any need to rely on a frame or a coefficient ring. As another example, we will see that a ring embedding of an abstract matrix ring can be based on the correspondence of two decomposition systems and the underlying rings.

Decomposition Systems of Modules

Definition 4.3.1. Decomposition system

Let $M := M_S$ be a module and $I = \{i : 0 \leq i < n + k\}$ an index set, where $n < \omega$ and $k \leq \omega$. A *decomposition system ε of M of format (n, k)* consists of

1. a decomposition $M = \bigoplus_{i \in I} M_i$ of M into a direct sum of submodules,

2. corresponding projections $\pi_i : M \twoheadrightarrow M_i$ and embeddings $\epsilon_i : M_i \hookrightarrow M$,

3. a family $\{\epsilon_{ij} : i, j \in I\}$ of maps $\epsilon_{ij} :: M_j \to M_i$.

4. submodules z_{ij} of M for $i \in I, j < n$, and

5. a 1-subring $C \leq End(M_0)$

such that the following conditions are satisfied.

1. For $i = j$, we have $\epsilon_{ii} = id_{M_i}$.

2. For $i, j < n$, ϵ_{ij} and ϵ_{ji} are mutually inverse morphisms, i.e., $\epsilon_{ij} \circ \epsilon_{ji} = id_{M_i}$.

3. For $i \in I$, we have $\epsilon_{i0} \circ \epsilon_{0i} = id_{M_i}$ (in particular, ϵ_{0i} is injective).

4. For distinct indices i, j, k such that $k, j < n$, we have $\epsilon_{ki} = \epsilon_{kj} \circ \epsilon_{ji}$

5. For $j < n$, z_{ij} is a relative complement of $im(\epsilon_{ji})$ in $[0, M_j]$.

6. For $i \in I$, $\epsilon_{0i} \circ \epsilon_{i0} \in C$.

In other words, for $i, j < n$, the submodules M_i, M_j are isomorphic, while for $i \in I, j < n$, M_i is isomorphic to a submodule of M_j – and the morphisms ϵ_{ji} are the corresponding isomorphisms and embeddings.

The relative complements z_{ij} are integrated into the notion of a decomposition systems for the following reason: For indices i, j with $j < n$, the injective morphism $\epsilon_{ji} : M_i \hookrightarrow M_j$ has a left inverse $\epsilon_{ij} :: M_j \to M_i$, defined only on $im(\epsilon_{ji}) \leq M_j$. Taking the relative complement $z_{ij} \leq M_j$ of $im(\epsilon_{ji})$ in $[0, M_j]$, we can extend the partial morphism $\epsilon_{ij} :: M_j \to M_i$ to a morphism $\epsilon_{ij} : M_j \to M_i$ by setting $\epsilon_{ij}(x) := 0$ for all $x \in z_{ij}$ (i.e., the extension $\epsilon_{ij} : M_j \to M_i$ acts trivially on z_{ij}).

Remark 4.3.1. For the ease of notation, we stated that a decomposition system contains a family of partial maps $\epsilon_{ij} :: M_j \to M_i, i, j \in I$. The required conditions should have made clear that in fact, we consider mainly morphisms $\epsilon_{ij} : M_j \to M_i$, but only particular maps have to exist – and those satisfy the desired relations. But one might take the view that all maps are partial maps, where the maps that are not important are considered to be partial maps with trivial domain.

Remark 4.3.2. We write $\varepsilon = \varepsilon(C, M)$ to indicate the ring C and the module M under consideration.

We recall Observation 4.1.2 for the natural identifications and conventions.

Definition 4.3.2. Morphisms of Dec. Systems
Let $M_S, M'_{S'}$ be modules over S, S' and $\varepsilon, \varepsilon'$ decomposition systems of M, M', respectively.[1] A morphism between the two decomposition systems $\varepsilon, \varepsilon'$ is a map $\eta : \varepsilon \to \varepsilon'$ such that the components of ε get mapped onto the components of ε'. In particular, the following hold.

1. $\eta(M_i) = M'_i$ and $\eta(\pi_i) = \pi'_i$, $\eta(\epsilon_i) = \epsilon'_i$ for all $i \in I$.

2. $\eta(z_{ij}) = z'_{ij}$ for all $i, j \in I$.

3. $\eta(\epsilon_{ij}) = \epsilon'_{ij}$ for all $i, j \in I$.

4. $\eta : C \to C'$ is a morphism of rings with units.

A morphism η between decomposition systems will be called *injective* or an *embedding of decomposition systems* if $\eta : C \to C'$ is injective.

The Abstract Matrix Ring of a Dec. System

Definition 4.3.3. Abstract matrix ring
Let M_S be a module and ε a decomposition system of M. The *abstract matrix ring* with respect to the decomposition system ε of M is

$$R(\varepsilon, C, M) := \{\varphi \in End(M_S) :$$
$$\epsilon_{0j} \circ \varphi_{ji} \circ \epsilon_{i0} \in C \text{ for all } i, j\} \subseteq End(M_S),$$

---

[1] Similarly, we denote the components of the two systems by the same letters, once with prime, once without.

where, as above, $\varphi_{ji} = \pi_j \circ \varphi \circ \epsilon_i$ and π_j, ϵ_i are the natural projections and embeddings belonging to decomposition system ε.

The following result justifies this definition.

Proposition 4.3.4. *The set $R(\varepsilon, C, M)$ is a 1-subring of $End(M_S)$.*

Remark 4.3.3. Note that the ring C occurring in the notation and the definition of an abstract matrix ring is part of the decomposition system ε.

▶ **Proof.** Firstly, we consider the identity map 1_M and the zero map 0_M. We distinguish the cases of equal and of different indices separately. For $i = j$ in I, we have

$$\epsilon_{0i} \circ 1_{ii} \circ \epsilon_{i0} = \epsilon_{0i} \circ \epsilon_{i0} \in C$$

and

$$\epsilon_{0i} \circ 0_{ii} \circ \epsilon_{i0} = 0_{|M_0} = 0 \in C.$$

For $i \neq j$ in I, we have

$$\epsilon_{0j} \circ 1_{ji} \circ \epsilon_{i0} = \epsilon_{0j} \circ 0_{ji} \circ \epsilon_{i0} = 0_{|M_0} = 0 \in C.$$

In particular, $R(\varepsilon, C, M)$ contains a unit.

Secondly, for $\varphi, \psi \in R(\varepsilon, C, M)$, we have

$$\epsilon_{0j} \circ (\varphi - \psi) \circ \epsilon_{i0} = \epsilon_{0j} \circ \varphi_{ji} \circ \epsilon_{i0} - \epsilon_{0j} \circ \psi_{ji} \circ \epsilon_{i0} \in C.$$

Therefore, $R(\varepsilon, C, M)$ contains additive inverses and is closed under addition.

Finally, for $\varphi, \psi \in R(\varepsilon, C, M)$, we have

$$\epsilon_{0j} \circ (\varphi \circ \psi)_{ji} \circ \epsilon_{i0} = \epsilon_{0j} \circ \left[\sum_k \varphi_{jk} \circ \psi_{ki} \right] \circ \epsilon_{i0}$$

$$= \epsilon_{0j} \circ \left[\sum_k \varphi_{jk} \circ (\epsilon_{k0} \circ \epsilon_{0k}) \circ \psi_{ki} \right] \circ \epsilon_{i0}$$

$$= \sum_k (\epsilon_{0j} \circ \varphi_{jk} \circ \epsilon_{k0}) \circ (\epsilon_{0k} \circ \psi_{ki} \circ \epsilon_{i0}).$$

Since $\epsilon_{0j} \varphi_{jk} \epsilon_{k0}$ and $\epsilon_{0k} \psi_{ki} \epsilon_{i0}$ are assumed to be elements of C and since C is a (sub)ring, we have

$$\epsilon_{0j} \circ (\varphi \circ \psi)_{ji} \circ \epsilon_{i0} \in R(\varepsilon, C, M).$$

◄

Dec. Systems and Ring Morphisms

Proposition 4.3.5. *Let* $M_S, M'_{S'}$ *be two modules with decomposition systems* $\varepsilon, \varepsilon'$ *and* $\eta : \varepsilon \to \varepsilon'$ *a morphism of decomposition systems between* ε *and* ε'. *Declaring*

$$\eta(\varphi_{ji}) := \epsilon'_{j0} \circ \eta \Big(\epsilon_{0j} \circ \varphi_{ji} \circ \epsilon_{i0} \Big) \circ \epsilon'_{0i}$$

for morphisms $\varphi_{ji} : M_i \to M_j$, *the map* η *can be extended to a map*

$$\eta : R(\varepsilon, C, M) \to R(\varepsilon', C', M')$$

in the following way. Since $\varphi \in End(M)$ *decomposes into* $\varphi = \bigoplus \varphi_i = \sum \varphi_{ji}$, *we can define* $\eta(\varphi_i) := \sum_{j \in I} \eta(\varphi_{ji})$

for a fixed index $i \in I$ and

$$\eta(\varphi) := \sum_{i \in I} \eta(\varphi_i) = \sum_{i,j \in I} \eta(\varphi_{ji}).$$

With this definition, $\eta : R(\varepsilon, C, M) \to R(\varepsilon', C', M')$ is a morphism of rings with unit.

If the restriction $\eta_{|_C} : C \to C'$ is injective, then so is the map $\eta : R(\varepsilon, C, M) \to R(\varepsilon', C', M')$.

▶ **Proof.** For the identity map $1 \in R$, we have $1 = \sum 1_{ii}$ and so

$$\eta(1) = \sum \eta(1_{ii}) = \sum \epsilon'_{i0} \circ \eta\left(\epsilon_{0i} \circ 1_{ii} \circ \epsilon_{i0}\right) \circ \epsilon'_{0i}$$

$$= \sum \epsilon'_{i0} \circ \eta\left(\epsilon_{0i} \circ \epsilon_{i0}\right) \circ \epsilon'_{0i} = \sum \epsilon'_{i0} \circ \epsilon'_{0i} \circ \epsilon'_{i0} \circ \epsilon'_{0i}$$

$$= \sum \epsilon'_{ii} = \sum id_{M'_i} = \sum 1'_{ii} = 1'.$$

Similarly, taking the decomposition $0 = \sum 0_{ji}$ of the zero map $0 \in R$, we get $\eta(0) = 0'$.

For addition of morphisms between submodules, consider $\varphi, \psi \in End(M_S)$ such that $\epsilon_{0j}\varphi_{ji}\epsilon_{i0}$ and $\epsilon_{0j}\psi_{ji}\epsilon_{i0}$ are elements of C for all indices i, j. Then

$$\eta(\varphi_{ji} + \psi_{ji}) = \epsilon^N_{j0} \circ \eta\left(\epsilon^M_{0j} \circ (\varphi_{ji} + \psi_{ji}) \circ \epsilon^M_{i0}\right) \circ \epsilon^N_{0i}$$

$$= \epsilon^N_{j0} \circ \eta\left(\epsilon^M_{0j} \circ \varphi_{ji} \circ \epsilon^M_{i0} + \epsilon^M_{0j} \circ \psi_{ji} \circ \epsilon^M_{i0}\right) \circ \epsilon^N_{0i}$$

$$= \epsilon^N_{j0} \circ \left(\eta\left(\epsilon^M_{0j}\varphi_{ji}\epsilon^M_{i0}\right) + \eta\left(\epsilon^M_{0j}\psi_{ji}\epsilon^M_{i0}\right)\right) \circ \epsilon^N_{0i}$$

$$= \epsilon^N_{j0} \circ \eta\left(\epsilon^M_{0j}\varphi_{ji}\epsilon^M_{i0}\right) \circ \epsilon^N_{0i} + \epsilon^N_{j0} \circ \eta\left(\epsilon^M_{0j}\psi_{ji}\epsilon^M_{i0}\right) \circ \epsilon^N_{0i}$$

$$= \eta(\varphi_{ji}) + \eta(\psi_{ji}).$$

Here, we have made use of the definition of η (first equality) and the fact that $\eta_{|C^M}$ is a ring homomorphism (third equality).

For composition of morphisms between submodules, consider $\varphi, \psi \in End(M_S)$ such that $\epsilon_{0j}\varphi_{ji}\epsilon_{i0}$ and $\epsilon_{0j}\psi_{ji}\epsilon_{i0}$ are elements of C for all indices i, j. Then

$$\eta(\varphi_{kj} \circ \psi_{ji})$$

$$\overset{\text{by def}}{=} \quad \epsilon'_{k0} \circ \eta\big(\epsilon_{0k} \circ (\varphi_{kj} \circ \psi_{ji}) \circ \epsilon_{i0}\big) \circ \epsilon'_{0i}$$

$$\overset{\text{ins. } id_{M_j}}{=} \quad \epsilon'_{k0} \circ \eta\big(\epsilon_{0k} \circ (\varphi_{kj} \circ \epsilon_{j0} \circ \epsilon_{0j} \circ \psi_{ji}) \circ \epsilon_{i0}\big) \circ \epsilon'_{0i}$$

$$\overset{\text{by assoc.}}{=} \quad \epsilon'_{k0} \circ \eta\Big((\epsilon_{0k}\varphi_{kj}\epsilon_{j0}) \circ (\epsilon_{0j}\psi_{ji}\epsilon_{i0})\Big) \circ \epsilon'_{0i}$$

$$\overset{\eta_{|C} \text{ morph.}}{=} \quad \epsilon'_{k0} \circ \eta(\epsilon_{0k}\varphi_{kj}\epsilon_{j0}) \circ \eta(\epsilon_{0j}\psi_{ji}\epsilon_{i0}) \circ \epsilon'_{0i}$$

$$\overset{\text{ins. } id_{N_j}}{=} \quad \epsilon'_{k0} \circ \eta(\epsilon_{0k}\varphi_{kj}\epsilon_{j0}) \circ (\epsilon'_{0j}\epsilon'_{j0}) \circ \eta(\epsilon_{0j}\psi_{ji}\epsilon_{i0})) \circ \epsilon'_{0i}$$

$$\overset{\text{by assoc.}}{=} \quad \big(\epsilon'_{k0}\eta(\epsilon_{0k}\varphi_{kj}\epsilon_{j0})\epsilon'_{0j}\big) \circ \big(\epsilon'_{j0}\eta(\epsilon_{0j}\psi_{ji}\epsilon_{i0})\epsilon'_{0i}\big)$$

$$= \quad \eta(\varphi_{kj}) \circ \eta(\psi_{ji}).$$

Now, let φ, ψ be in $R(\varepsilon, C, M)$.

For addition, we have $(\varphi + \psi)_{ji} = \varphi_{ji} + \psi_{ji}$. It follows that

$$\begin{aligned}
\eta(\varphi + \psi) &= \sum \eta((\varphi + \psi)_{ji}) = \sum \eta(\varphi_{ji} + \psi_{ji}) \\
&= \sum \eta(\varphi_{ji}) + \eta(\psi_{ji}) = \sum \eta(\varphi_{ji}) + \sum \eta(\psi_{ji}) \\
&= \eta(\varphi) + \eta(\psi).
\end{aligned}$$

For multiplication, we have $(\varphi \circ \psi)_{ki} = \sum_j \varphi_{kj} \circ \psi_{ji}$.

Hence,

$$
\begin{aligned}
\eta(\varphi \circ \psi) &= \sum_{i,k} \eta\Big(\sum_j \varphi_{kj} \circ \psi_{ji}\Big) = \sum_{i,k}\sum_j \eta(\varphi_{kj} \circ \psi_{ji}) \\
&= \sum_{i,k}\sum_j \eta(\varphi_{kj}) \circ \eta(\psi_{ji}) = \sum_{i,j,k} \eta(\varphi_{kj}) \circ \eta(\psi_{ji}).
\end{aligned}
$$

On the other hand, we have

$$
\eta(\varphi) \circ \eta(\psi) = \sum_{k,j} \eta(\varphi_{kj}) \circ \sum_{j,i} \eta(\psi_{ji}) = \sum_{i,j,k} \eta(\varphi_{kj}) \circ \eta(\psi_{ji}).
$$

Thus, $\eta(\varphi \circ \psi) = \eta(\varphi) \circ \eta(\psi)$.

Finally, assume that $\eta_{|C} : C \to C'$ is injective. Let $\varphi \in R(\varepsilon, C, M)$ and assume that $\eta(\varphi) = 0$. Since M' decomposes into a direct sum $M' = \bigoplus M'_i$, this implies $\eta(\varphi_{ji}) = 0$ for all $i, j \in I$. But then, we have

$$
0 = \epsilon'_{0j} \circ \eta(\varphi_{ji}) \circ \epsilon'_{i0} = \eta(\epsilon_{0j} \circ \varphi_{ji} \circ \epsilon_{i0}) \quad \text{for all } i, j \in I.
$$

Since $\eta_{|C} : C \to C'$ is injective, we have $\epsilon_{0j} \circ \varphi_{ji} \circ \epsilon_{i0} = 0$ for all $i, j \in I$. Since $\epsilon_{i0} : M_0 \to M_i$ is surjective and $\epsilon_{0j} : M_j \to M_0$ is injective, we have $\varphi_{ji} = 0$ for all $i, j \in I$, hence $\varphi = 0$. ◄

4.4 Frames and Induced Structures

In this section, we approach the connection between the general framework and our particular setting. Starting with a frame in $L(M)$, we will develop the notion of the coefficient ring of a frame and the notion of an induced decomposition system. Still, we stay one level of abstraction away from our final setting of $*$-regular rings.

The Coefficient Ring of a Frame

Definition 4.4.1. Coefficient ring of a frame
Let M_S be a right module over S and Φ a stable (n, k)-frame in $L(M_S)$, contained in the sublattice $L \leq L(M_S)$ with $n \geq 3$. The *coefficient ring* of (Φ, L, M) is

$$C(\Phi, L, M) := \{\varphi \in End(M_0) : \Gamma(\epsilon_{10} \circ \varphi) \in L\},$$

a subset of $End(M_0)$.

The following result justifies this definition.

Proposition 4.4.2. *The set $C(\Phi, L, M)$ is a 1-subring of $End(M_0)$.*

▶ **Proof.** The line of arguments and the technique of this proof are well-known: It is possible to express the ring operations via lattice terms with constants in Φ. (See the works of von Neumann, Jónsson and Handelman.) These terms are uniform in the frame Φ. In particular, they are independent of the particular module M_S. Nevertheless, we will present the approach in this section.

For the nullary ring operations, consider the identity map

$$1 := 1_{M_0} : M_0 \to M_0$$

and the zero map

$$0 := 0_{M_0} : M_0 \to M_0.$$

We have $\epsilon_{10} \circ 1 = \epsilon_{10}$ and

$$\Gamma(\epsilon_{10}) = a_{10} \in \Phi \subseteq L,$$

so $1 \in C$. Likewise, we have $\epsilon_{10} \circ 0 = 0_{10}$ and

$$\Gamma(0_{10}) = \{x - 0(x) : x \in M_0\} = a_0 \in \Phi \subseteq L,$$

so $0 \in C$.

It remains to consider the unary and binary ring operations. Some preliminary work is needed.

Lemma 4.4.3. *We define the following auxiliary terms.*

$$S(Z) := \big((Z + a_2) \cdot [a_{20} + a_1] + a_0\big) \cdot [a_1 + a_2]$$

$$S^{201}(Z) := (Z + a_{20}) \cdot [a_2 + a_1]$$
$$S^{210}(Z) := (Z + a_{21}) \cdot [a_2 + a_0]$$

Then the following hold for a morphism $\psi : M_0 \to M_1$ with graph $\Gamma(\psi) = \{x - \psi x : x \in M_0\}$.

$$S(\Gamma(\psi)) = \{-\psi y - \epsilon_{20} y : y \in M_0\} = \{\psi y + \epsilon_{20} y : y \in M_0\}$$

$$S^{201}(\Gamma(\psi)) = \{\epsilon_{20} z - \psi z : z \in M_0\}$$
$$S^{210}(\Gamma(\psi)) = \{z - \epsilon_{21} \psi z : z \in M_0\}$$

In particular, for $\psi = \epsilon_{10} \circ \varphi$ with $\varphi : M_0 \to M_0$, the application of one of the terms above yields the stated result.

▷ **Proof.** For $S(\Gamma(\psi))$, we get

$$(\Gamma(\psi) + a_2) \cdot [a_{20} + a_1]$$
$$= \{x - \psi x + \lambda : x \in M_0, \lambda \in M_2\}$$
$$\cdot \{z - \epsilon_{20} z + \mu : z \in M_0, \mu \in M_1\}$$
$$= \{z - \psi z - \epsilon_{20} z : z \in M_0\}$$

and

$$(\{z - \psi z - \epsilon_{20} z : z \in M_0\} + a_0) \cdot [a_1 + a_2]$$
$$= \{z - \psi z - \epsilon_{20} z + \nu : z, \nu \in M_0\} \cdot [a_1 + a_2]$$
$$= \{-\psi y - \epsilon_{20} y : y \in M_0\}.$$

For $S^{201}(\Gamma(\psi))$, we get

$$\{x - \psi x - z + \epsilon_{20} z : x, z \in M_0\} \cdot [a_1 + a_2]$$
$$= \{\epsilon_{20} z - \psi z : z \in M_0\}.$$

For $S^{210}(\Gamma(\psi))$, we get

$$\{x - \psi x + z - \epsilon_{21} z : x \in M_0, z \in M_1\} \cdot [a_0 + a_2]$$
$$= \{z - \epsilon_{21} \psi z : z \in M_1\}.$$

◁

Remark 4.4.1. For $\varphi \in C$, we agree to define its graph to be $\{x - \epsilon_{10} \circ \varphi x : x \in M_0\}$. To simplify calculations, we denote $\epsilon_{10} \circ \varphi$ by ψ. In particular, we will do so when dealing with additive inverses and addition. Note that composition requires separate handling: We can compose $\varphi_1, \varphi_2 : M_0 \to M_0$, but not $\psi_1, \psi_2 : M_0 \to M_1$.

Now, we consider the remaining ring operations again. For negation, we define the following lattice term.

$$F_-(X) := (S(X) + a_{02}) \cdot [a_0 + a_1]$$

Using the previous result concerning $S(X)$, we get

$$
\begin{aligned}
& F_-(\Gamma(\psi)) \\
= & \left(\{-\psi y - \epsilon_{20} y : y \in M_0\} + a_{02}\right) \cdot [a_0 + a_1] \\
= & \{-\psi y - \epsilon_{20} y + z - \epsilon_{02} z : y \in M_0, z \in M_2\} \cdot [a_0 + a_1] \\
= & \{-\psi y - \epsilon_{02}\epsilon_{20} y : y \in M_0\} = \{-\psi y - y : y \in M_0\} \\
= & \Gamma(-\psi).
\end{aligned}
$$

For addition, we define the following lattice term.

$$F_+(U, V) := \Big((U + S(V)) \cdot [a_{20} + a_1] + a_2\Big) \cdot [a_0 + a_1]$$

If we plug in $\Gamma(\varphi)$ for U and $\Gamma(\psi)$ for V, we get

$$
\begin{aligned}
F_+(\Gamma(\varphi), \Gamma(\psi)) = & \Big(\{x - \varphi x - \epsilon_{20} y - \psi y : x, y \in M_0\} \\
& \cdot [a_{20} + a_1] + a_2\Big) \cdot [a_0 + a_1].
\end{aligned}
$$

We calculate the big bracket first.

$$
\begin{aligned}
& \Big(\{x - \varphi x - \epsilon_{20} y - \psi y : x, y \in M_0\} \cdot [a_{20} + a_1] + a_2\Big) \\
= & \Big(\{x - \varphi x - \epsilon_{20} y - \psi y : x, y \in M_0\} \\
& \cdot \{z - \epsilon_{20} z + \lambda : z \in M_0, \lambda \in M_1\} + a_2\Big)
\end{aligned}
$$

We determine this intersection and take into account the next step of calculation. We arrive at

$$F_+(\Gamma(\varphi), \Gamma(\psi))$$
$$= \left(\{x - \varphi x - \epsilon_{20}x - \psi x : x \in M_0\} + a_2\right) \cdot [a_0 + a_1]$$
$$= \{x - \varphi x - \psi x : x \in M_0\} = \{x - (\varphi + \psi)x : x \in M_0\}$$
$$= \Gamma(\varphi + \psi).$$

For multiplication, we consider $\varphi, \psi : M_0 \to M_0$. As before, their graphs are defined via ϵ_{10} by

$$\Gamma(\varphi) := \{x - \epsilon_{10} \circ \varphi x : x \in M_0\}$$

and

$$\Gamma(\psi) := \{y - \epsilon_{10} \circ \psi y : y \in M_0\}.$$

We define the following lattice term.

$$F_\circ(U, V) := \left(S^{201}(U) + S^{210}(V)\right) \cdot [a_0 + a_1]$$

It follows that

$$F_\circ(\Gamma(\varphi), \Gamma(\psi))$$
$$= \left(\{\epsilon_{20}x - \epsilon_{10}\varphi x + y - \epsilon_{21}\epsilon_{10}\psi y : x, y \in M_0\}\right) \cdot [a_0 + a_1]$$
$$= \{y - \epsilon_{10}\varphi\psi y : y \in M_0\}$$
$$= \Gamma(\varphi \circ \psi),$$

since $\varphi \circ \psi : M_0 \to M_0$ and $\epsilon_{21}\epsilon_{10} = \epsilon_{20}$ is injective. In particular, the intersection with $[a_0 + a_1]$ forces that $\epsilon_{20}x = \epsilon_{20}\psi y$ and consequently $x = \psi y$. ◀

The Decomposition System of a Frame

Proposition 4.4.4. *Let M_S be a right module over S and Φ a stable (n, k)-frame in $L(M_S)$ contained in the sublattice $L \leq L(M_S)$ with $n \geq 3$.*

Then Φ induces a decomposition system in the following way. Since Φ is a frame, we have a decomposition of M into a direct sum $M = \bigoplus M_i$, together with corresponding projections and embeddings. As usual, the axes of perspectivity as well as the axes of subperspectivity are the graphs of morphisms between the summands. Since Φ is stable, we have relative complements z_{ij} as required.[2] As ring C, we take the coefficient ring $C(\Phi, L, M)$.

We denote the decomposition system induced by the frame Φ by $\xi = \xi_{\Phi, L}(C, M)$.

▶ **Proof.** The only thing left to show is Property 5 in Definition 4.3.1. Consider ϵ_{0i} and ϵ_{i0}. Both graphs $\Gamma(\epsilon_{0i}), \Gamma(\epsilon_{i0})$ and clearly $\Gamma(\epsilon_{10})$ lie in the frame Φ. Since we can express composition of maps by lattice terms with constants in Φ, we have $\Gamma(\epsilon_{10} \circ \epsilon_{0i} \circ \epsilon_{i0}) \in L$. ◄

The Matrix Ring of a Frame

In Section 4.3, we defined the abstract matrix ring of a given decomposition system $\varepsilon = \varepsilon(C, M)$. In Section 4.4, we introduced the coefficient ring $C(\Phi, L, M)$ of a given stable frame Φ. Consequently, we have the following.

Definition 4.4.5. Matrix ring of a frame
Let M_S be module, Φ a stable (n, k)-frame in $L(M_S)$,

---

[2]That is, z_{ij} a relative complement of b_{ji} in $[0, a_j]$, where b_{ji} is the image of a_i under a_{ji} in $[0, a_j]$.

with $n \geq 3$, contained in the sublattice $L \leq L(M_S)$, $C(\Phi, L, M)$ the coefficient ring as defined in Section 4.4, and $\xi = \xi_{\Phi,L}(C, M)$ the induced decomposition system as defined in Section 4.4. The ring

$$R(\Phi, L, M) := R(\xi, C(\Phi, L, M), M)$$

will be called the *matrix ring* (of Φ, L, M).

Frames and Morphisms

Let $M = M_S$ and $M' = M'_{S'}$ be modules, $L \leq L(M_S)$ a complemented 0-1-sublattice, and Φ a stable frame in $L(M_S)$ contained in L of format (n, k) with $n \geq 3$. Assume that we are given a morphism $\iota : L \hookrightarrow L(M')$ of bounded complemented lattices.

Observation 4.4.1. The image $\Phi' := \iota[\Phi]$ is a stable frame in $L(M')$, contained in the sublattice $L' := \iota[L]$ of $L(M')$. In particular, we have $\iota(M_i) = M'_i$, $\iota(\pi_i) = \pi'_i$, $\iota(\epsilon_i) = \epsilon'_i$ and $\iota(z_{ij}) = z_{ij}$.

Proposition 4.4.6. *The morphism $\iota : L \to L'$ induces a morphism η between the induced decomposition systems*

$$\xi := \xi_{\Phi,L}(C(\Phi, L, M), L, M)$$

and

$$\xi' := \xi'_{\Phi',L'}(C(\Phi', L', M), L', M').$$

If $\iota : L \to L'$ is injective, then so is $\eta : \xi \to \xi'$.

▶ **Proof.** We want to define η via the lattice morphism $\iota : L \to L'$. For the first two properties of a morphism between two decomposition systems (see Definition 4.3.2), we define η to coincide with ι on the submodules M_i, z_{ij} of M and recall Observation 4.4.1.

Now consider the morphisms ϵ_{ji} given by the frame Φ, i.e., $\Gamma(\epsilon_{ji}) = a_{ji} \in \Phi$. Then

$$\iota\big(\Gamma(\epsilon_{ji})\big) = \iota(a_{ji}) = a'_{ji} = \Gamma(\epsilon'_{ji}) \in \Phi'.$$

Setting

$$\eta(\epsilon_{ji}) := \epsilon'_{ji} \text{ for } i \neq j < n$$

and

$$\eta(\epsilon_{ii}) := \epsilon'_{ii} \text{ for arbitrary } i,$$

we have guaranteed that η maps the morphism ϵ_{ji} to ϵ'_{ji}.

For a triple (i, j, k) of appropriate indices, the compatibility $\epsilon_{ki} = \epsilon_{kj} \circ \epsilon_{ji}$ of maps is determined by the lattice-theoretical equation

$$[a_{kj} + a_{ji}] \cdot [a_k + a_i] = a_{ki}$$

of the elements of the frame Φ (and similarly for Φ'). Therefore, we have

$$\eta(\epsilon_{kj} \circ \epsilon_{ji}) = \eta(\epsilon_{ki}) = \epsilon'_{ki} = \epsilon'_{kj} \circ \epsilon'_{ji}$$

for appropriate indices i, j, k.

Secondly, we consider an element φ of the coefficient ring $C = C(\Phi, L, M)$, i.e., a morphism $\varphi : M_0 \to M_0$ with $\Gamma(\epsilon_{10} \circ \varphi) \in L$.

The property that $\epsilon_{10} \circ \varphi$ is a morphism between M_0 and M_1 is equivalent to the lattice-theoretical property that $\Gamma(\epsilon_{10} \circ \varphi)$ is a relative complement of M_1 in $[0, M_0 + M_1]$. Since $\iota : L \to L'$ is a lattice morphism mapping Φ

to Φ', $\iota(\Gamma(\epsilon_{10} \circ \varphi))$ is a relative complement of M_1' in $[0, M_0' + M_1']$, i.e., the graph of a morphism $\psi : M_0' \to M_1'$. Composing ψ with ϵ_{01}', we can define

$$\eta(\varphi) := \epsilon_{01}' \circ \psi : M_0' \to M_0'.$$

Thirdly, we can capture the ring operations on $C(\Phi, L, M)$ via lattice terms with constants in Φ. Hence, the ring operations are transferred via $\iota : L \to L'$ to Φ' and C'. Accordingly, the map

$$\eta : C(\Phi, L, M) \to C(\Phi', L', M')$$

is a morphism of rings.

Finally, injectivity of ι implies injectivity of η. ◀

Corollary 4.4.7. *In the given situation, there exists a morphism*

$$\eta : R(\Phi, L, C) \to R(\iota[\Phi], \iota[L], C')$$

of rings with unit.
If $\iota : L \to L'$ is injective, then so is $\eta : R \to R'$.

In particular, if L embedds into the subspace lattice $L(V)$ of a vector space V, we have a ring embedding

$$\eta : R(\Phi, L, C^M) \hookrightarrow End(V_D)$$

▷ **Proof.** Combine Proposition 4.4.6 and Proposition 4.3.5. ◁

4.5 Representability of ∗-Reg. Rings

In this section, we will finally deal with ∗-regular rings. First, we focus our attention on a ∗-regular ring R with unit such that the MOL $\overline{L}(R_R)$ contains a stable orthogonal frame. With that restriction, we aim at representability of simple ∗-regular rings with unit. Subsequently, we will deal with simple ∗-regular rings without unit and finally, with subdirectly irreducible ∗-regular rings (with and without unit).

We recall the following facts. Since the class $\mathcal{R}$ of all ∗-regular rings is a variety, we can and will focus our attention on *subdirectly irreducible* ∗-regular rings. Due to Theorem 1.7.4 and its Corollary 1.7.5, we can concentrate on *countable subdirectly irreducible* ∗-regular rings.

Furthermore, we bear in mind that every subdirectly irreducible Artinian ∗-regular ring is simple (see Lemma 1.3.15) and that every subdirectly irreducible Artinian ∗-regular ring has a faithful linear representation (see Lemma 1.6.12). In particular, a subdirectly irreducible ∗-regular ring R of heigth $L(R) \leq 2$ has a faithful representation.

∗-Regular Rings with Frames

Remark 4.5.1. For this section, we assume the following.

1. R is a ∗-regular ring with unit.

2. $L := \overline{L}(R_R)$ is a MOL of height $h(L) \geq 3$.

3. Φ is a stable orthogonal (n, k)-frame in L, $n = 3$.

4. $M_R = R_R$, if not stated otherwise.

Remark 4.5.2. In particular, by Theorem 1.6.9, there exists a faithful representation $\iota : L \hookrightarrow L(V_D, \langle \cdot, \cdot \rangle)$ of the MOL L. Consequently, Corollary 4.4.7 applies in its full strength.

The MOL $\overline{L}(R_R)$, Frames and Dec. Systems

Corollary 4.5.1. *Let e_i and e_j be projections in R with the corresponding cyclic modules $e_i R$ and $e_j R$.*

Any morphism $\varphi_{ji} : e_i R \to e_j R$ is a left multiplication by a ring element $e_j s e_i \in e_j R e_i$.

▷ **Proof.** See also Observation 4.1.3. The action of φ_{ji} on $e_i R$ is defined by the action of φ_{ji} on e_i. We have

$$\varphi_{ji}(e_i) = \varphi_{ji}(e_i^2) = \varphi_{ji}(e_i)e_i = ae_i \qquad \text{for some } a \in R,$$

that is, $\varphi_{ji} = \widehat{ae_i}$. Since $ae_i \in e_j R$, we have $ae_i = e_j s$ for some $s \in R$. But then $ae_i = ae_i^2 = e_j s e_i$, hence $\varphi_{ji} = \widehat{e_i s e_j}$. ◁

Observation 4.5.1. By Proposition 4.4.4, the stable orthogonal frame Φ induces a decomposition system

$$\xi = \xi_{\Phi,L}(C, M) = \xi_{\Phi,L}(C, R_R).$$

More exactly, we have the following correspondences.

1. The summands M_i correspond to principal right ideals $e_i R$ generated by a projection e_i. Each projection π_i corresponds to a map $\widehat{e_i}$ given by left multiplication by e_i and coincides with the (extension of the) embedding $\epsilon_i = id_{M_i}$.

2. The morphisms $\epsilon_{ji} : e_i R \to e_j R$ are given by $\widehat{e}_{ji}$
   with $e_{ji} := \epsilon_{ji}(e_i) \in e_i Re_j$.

3. For the coefficient ring of the frame, we have

$$C = C(\Phi, L, M) = C(\Phi, L, R_R) = \{\widehat{r} : r \in e_0 Re_0\}.$$

We prove the last statement.

▷ **Proof.** We recall the definition of the coefficient ring
of a frame:

$$C(\Phi, L, M) = \{\varphi \in End(M_0) : \Gamma(\epsilon_{10} \circ \varphi) \in L\}.$$

In the given situation, we have $M_0 = e_0 R$ for a projection
e_0, hence an endomorphism $\varphi \in End(M_0)$ is given by left
multiplication $\widehat{a}$ for some $a \in e_0 Re_0$. On the other hand,
each element a of $e_0 Re_0$ gives rise to an endomorphism
$\widehat{a} \in End(M_0)$. Finally, for $a \in e_0 Re_0$, we have

$$\Gamma(\epsilon_{10} \circ \widehat{a}) = \Gamma(\widehat{e_{10} \cdot a}) = (e_0 - e_{10}a)R = pR$$

for some $p \in P(R)$. Hence, $\Gamma(\epsilon_{10} \circ \widehat{a}) \in L$, proving that
$C(\Phi, L, M) = End(M_0) = \{\widehat{r} : r \in e_0 Re_0\}$. ◁

Remark 4.5.3. From now on, we will denote the mor-
phisms ϵ_{ji} given by the decomposition above by ϵ_{ji} and
$\widehat{e}_{ji}$ interchangedly, as it suits the particular situation.

Corollary 4.5.2. *We have an isomorphism*

$$\theta : R \to R(\Phi, L, R_R)$$

of rings with unit.

▷ **Proof.** As stated in Remark 4.5.1 at the beginning of the section, we agree to write M for R_R. We recall the definition of the matrix ring of a frame:

$$R(\Phi, L, M) = \{\varphi \in End(M) :$$
$$\forall i, j \in I.\ \epsilon_{0j} \circ \varphi_{ji} \circ \epsilon_{i0} \in C(\Phi, L, M)\}.$$

Since R contains a unit, an endomorphism $\varphi \in End(M)$ is given by left multiplication $\hat{r}$ for some $r \in R$. We notice that

$$(\hat{r})_{ji} = \pi_j \circ \hat{r} \circ \epsilon_i = \hat{e}_j \circ \hat{r} \circ \hat{e}_i = \widehat{e_j r e_i} = \widehat{r_{ji}}$$

with $r_{ji} := e_j r e_i$. Then we have

$$\epsilon_{0j} \circ (\hat{r})_{ji} \circ \epsilon_{i0} = \hat{e}_{0j} \circ \widehat{e_j r e_i} \circ \hat{e}_{i0} = \widehat{e_{0j} r e_{i0}},$$

since $e_{0j} e_j = e_{0j}$ and $e_i e_{i0} = e_{i0}$.

The equality $\epsilon_{0j} \circ (\hat{r})_{ji} \circ \epsilon_{i0} = \widehat{e_{0j} r e_{i0}}$ and Observation 4.5.1 lead to $\epsilon_{0j} \circ (\hat{r})_{ji} \circ \epsilon_{i0} \in C(\Phi, L, M)$. Since the indices i, j have been arbitrary, we can conclude that $\hat{r} \in R(\Phi, L, M)$.

In particular, for an element $r \in R$, the left multiplication $\hat{r} : M \to M$ decomposes into

$$\hat{r} = \sum \widehat{r_{ji}}, \qquad \text{where}$$

$$\widehat{r_{ji}} : e_i R \to e_j R \quad \text{and} \quad r_{ji} = e_j r e_i \in e_j R e_i.$$

That is, the isomorphism $\theta : R \to R(\Phi, L, M)$ is given by $\theta : r \mapsto \hat{r}$.

Of course, we have

$$\begin{aligned}
\Gamma(\epsilon_{10} \circ \widehat{e_{0j}re_{i0}}) &= \Gamma(\widehat{e_{10}} \circ \widehat{e_{0j}re_{i0}}) = \Gamma(\widehat{e_{1j}re_{i0}}) \\
&= (e_0 - e_{1j}re_{i0})R \in L = \overline{L}(R_R).
\end{aligned}$$

◁

Corollary 4.5.3. *We have an embedding*

$$\rho : R \hookrightarrow End(V_D)$$

of rings with unit.

▷ Proof. By Remark 4.5.2 and Theorem 1.6.9, the MOL $L = \overline{L}(R_R)$ is representable in $L(V_D, \langle \cdot, \cdot \rangle)$. We combine that with Corollaries 4.4.7 and 4.5.2 and define $\rho := \eta \circ \theta$ to get the desired isomorphism. ◁

Involution on R and Adjointness on $End(V_D, \langle ., . \rangle)$

In this section, we consider the involution on R. In particular, we want to stress similarities between the involution on R and the adjointness on V_D.

Remark 4.5.4. For the present section, we assume the following.

1. $(V_D, \langle \cdot, \cdot \rangle)$ is a vector with scalar product.

2. K is a MOL represented in $L(V_D, \langle \cdot, \cdot \rangle)$, i.e., K is a modular sublattice of $L(V_D)$ and the orthocomplementation on K is induced by the scalar product $\langle \cdot, \cdot \rangle$ on V_D.

3. Ψ is a stable orthogonal frame in the MOL K.

4. U_i and U_j are subspaces of V with $U_i, U_j \in \Psi$ and $f : U_i \rightarrow U_j, g : U_j \rightarrow U_i$ are linear maps.

5. The usual conventions on R, M, L, Φ hold.

If we discuss both situations simultaneously, we use the symbol Υ for the frame, N_i for submodules or subspaces and a, b for morphisms.

Definition 4.5.4. Adjointness on $End(V_D, \langle ., . \rangle)$
We call f and g *adjoint to each other* (with respect to $\langle \cdot, \cdot \rangle$) if

$$\forall v \in U_i, w \in U_j. \qquad \langle fv, w \rangle = \langle v, gw \rangle$$

Remark 4.5.5. Notice that U_i and U_j are elements of the orthogonal frame Ψ, in particular, if $i \neq j$, then U_i and U_j are orthogonal to each other.

Lemma 4.5.5. *The following conditions are equivalent.*

1. *f and g are adjoint to each other in the sense of Definition 4.5.4.*

2. *The extensions $\overline{f}, \overline{g} : V \rightarrow V$ are adjoint to each other in the usual sense.*

If $i \neq j$, both conditions are equivalent to $\Gamma(f) \perp \Gamma(-g)$.[3]

▷ Proof. The equivalence of the first two conditions is immediate. Now, if $i \neq j$, we have

$$\Gamma(f) = \{v - fv : v \in U_i\} \qquad \Gamma(-g) = \{w + gw : w \in U_j\}$$

---

[3]Note that $\Gamma(f), \Gamma(g), \Gamma(-g)$ are contained in the MOL K.

and

$$\Gamma(f) \perp \Gamma(-g)$$

$\Leftrightarrow \quad \langle v - fv, w + gw \rangle = 0$ for all $v \in U_i, w \in U_j$

$\Leftrightarrow \quad \langle v, w \rangle + \langle v, gw \rangle - \langle fv, w \rangle - \langle fv, gw \rangle = 0$
   for all $v \in U_i, w \in U_j$

$\Leftrightarrow \quad \langle v, gw \rangle - \langle fv, w \rangle = 0$ for all $v \in U_i, w \in U_j$

$\Leftrightarrow \quad \langle v, gw \rangle = \langle fv, w \rangle$ for all $v \in U_i, w \in U_j$

$\Leftrightarrow \quad f$ and g are adjoint in the sense of Def. 4.5.4.

The terms $\langle v, w \rangle$ and $\langle fv, gw \rangle$ vanish because U_i and U_j are orthogonal to each other. ◁

Now, we derive a result similar to Lemma 4.5.5 for a *-regular ring R and the relation between the involution on R and the orthogonality on $L = \overline{L}(R_R)$.

Lemma 4.5.6. *The involution on R can be captured via the orthogonality on L. More exactly, for $a_{ij} \in e_i R e_j$ and $b_{ji} \in e_j R e_i$ with $i \neq j$, the following conditions are equivalent.*

1. $a_{ij} = b_{ji}^*$.

2. $\Gamma(\widehat{a_{ij}}) \perp \Gamma(-\widehat{b_{ji}})$.

▷ Proof. Since $\widehat{a_{ij}} : e_j R \to e_i R$ and $-\widehat{b_{ji}} : e_i R \to e_j$,

$$\Gamma(\widehat{a_{ij}}) = (e_j - a_{ij}e_j)R \quad \text{and} \quad \Gamma(-\widehat{b_{ji}}) = (e_i + b_{ji}e_i)R.$$

The orthogonality on L is given by $pR \perp qR :\Leftrightarrow q^*p = 0$.

Calculating yields

$$(e_i + b_{ji}e_i)^* \cdot (e_j - a_{ij}e_j) = (e_i + e_i b_{ji}^*)(e_j - a_{ij}e_j)$$
$$= e_i e_j - e_i a_{ij} e_j + e_i b_{ji}^* e_j - e_i b_{ji}^* a_{ij} e_j$$
$$= e_i(-a_{ij} + b_{ji}^*)e_j = -a_{ij} + b_{ji}^*,$$

since $a_{ij}, b_{ji}^* \in e_i R e_j$. Hence, $a_{ij} = b_{ji}^*$ iff $\Gamma(\widehat{a_{ij}}) \perp \Gamma(-\widehat{b_{ji}})$.
◁

Corollary 4.5.7. Uniqueness
Let (i, j) be an arbitrary pair of indices.

A linear map $f : U_i \to U_j$ has at most one adjoint $g : U_j \to U_i$. Due to this uniqueness, it is legitimate to write $f^* = g$ if f, g are adjoint to each other.

Likewise, a map $\widehat{a_{ij}} : e_j R \to e_i R$ gives rise to a map $\widehat{b_{ji}} : e_i R \to e_j R$, namely $\widehat{b_{ji}} = \widehat{a_{ij}^*}$. If $i \neq j$, we have $\widehat{b_{ji}} = \widehat{a_{ij}^*}$ iff $\Gamma(\widehat{a_{ij}}) \perp \Gamma(-\widehat{b_{ji}})$.

Lemma 4.5.8. *For each map $\epsilon_{ki} : N_i \to N_k$, there exists an adjoint $\epsilon_{ki}^* : N_k \to N_i$ in Υ.*

▷ Proof. Let $i \neq k$. $G := \Gamma(\epsilon_{ki})$ is a relative complement of M_k in $[0, M_i + M_k]$. We claim that

$$H := G^\perp \cap (M_i + M_k)$$

is the graph of a morphism $h : M_k \to M_i$.
▷ Proof. We have to show that H is a relative complement of M_i in $[0, M_i + M_k]$. We focus our attention on a lower section of the submodule lattice, considering the interval $[0, M_i + M_k]$ only. Then we have the following situation:

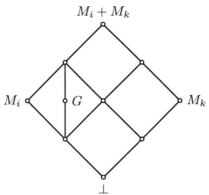

$$M_i + M_k$$

$$M_i \qquad G \qquad M_k$$

$$\perp$$

Taking complements, we get the following picture in the corresponding upper section of the submodule lattice:

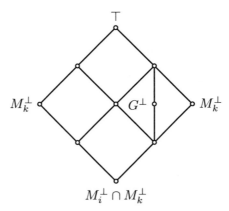

$$\top$$

$$M_k^\perp \qquad G^\perp \qquad M_k^\perp$$

$$M_i^\perp \cap M_k^\perp$$

Now we intersect with $(M_i + M_k)$ to pull this section down into the interval $[0, M_i + M_k]$ again. With $H := G^\perp \cap (M_i + M_k)$, we have the following picture.

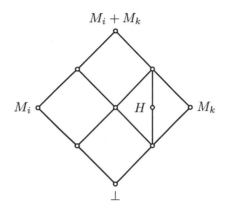

$$M_i + M_k$$

$$M_i \qquad H \qquad M_k$$

$$\bot$$

◁

Consequently, H is a relative complement of M_i in $[0, M_i + M_k]$. By the previous characterisation of adjointness (in both senses) via orthogonality of graphs, we have proven the desired existence of an adjoint $\epsilon_{ki}^* : N_k \to N_i$ of $\epsilon_{ki} : N_i \to N_k$.

Of course, for the situation of a ∗-regular ring, the existence of ϵ_{ki}^* can be concluded directly by $\epsilon_{ki}^* = \widehat{e_{ki}^*}$. ◁

Corollary 4.5.9. *For $i, k \in I$ with $k < n$, we have $\epsilon_{ki}^* \circ \epsilon_{ik}^* = id_{U_i}$.*

▷ Proof. We have

$$\epsilon_{ki}^* \circ \epsilon_{ik}^* = (\epsilon_{ik} \circ \epsilon_{ki})^* = (id_{N_i})^* = id_{N_i}.$$

◁

Remark 4.5.6. Obviously, Lemma 4.5.8 and Corollary 4.5.9 hold for arbitrary indices $i, k \in I$: Recall that for if $i = k$, we have $\epsilon_{ik} = \epsilon_{ii} = \epsilon_i = id_{N_i}$, which is an Hermitian idempotent map.

Corollary 4.5.10. *Let $a : N_i \to N_j$ and $b : N_j \to N_i$ be as before.*

Then a and b are adjoint to each other iff $\epsilon_{i0}^* \circ b \circ \epsilon_{1j}^*$ and $\epsilon_{1j} \circ a \circ \epsilon_{i0}$ are adjoint to each other.

▷ **Proof.** Assume that $\epsilon_{i0}^* \circ b \circ \epsilon_{1j}^*$ and $\epsilon_{1j} \circ a \circ \epsilon_{i0}$ are adjoint to each other, where adjoint is either understood in the sense of Definition 4.5.4 or in the sense of Lemma 4.5.6. Then

$$\epsilon_{0i}^* \circ (\epsilon_{1j} \circ a \circ \epsilon_{i0})^* \circ \epsilon_{j1}^* = (\epsilon_{j1}\epsilon_{1j}a\epsilon_{i0}\epsilon_{0i})^* = a^*$$

and

$$\begin{aligned} & \epsilon_{0i}^* \circ (\epsilon_{1j} \circ a \circ \epsilon_{i0})^* \circ \epsilon_{j1}^* \\ = \ & \epsilon_{0i}^* \circ (\epsilon_{i0}^* \circ b \circ \epsilon_{1j}^*) \circ \epsilon_{j1}^* \\ = \ & (\epsilon_{0i}^*\epsilon_{i0}^*)b(\epsilon_{1j}^*\epsilon_{j1}^*) = b, \end{aligned}$$

so $a^* = b$.

Now, assume that a, b are adjoint to each other. Then

$$(\epsilon_{1j} \circ a \circ \epsilon_{i0})^* = \epsilon_{i0}^* \circ a^* \circ \epsilon_{1j}^* = \epsilon_{i0}^* \circ b \circ \epsilon_{1j}^*.$$

◁

Remark 4.5.7. Corollary 4.5.10 holds for arbitrary indices i and j, too. In particular, we can complete Lemma 4.5.6 and Corollary 4.5.7 by noting that if $a_{ii}, b_{ii} \in e_i R e_i$, we have

$$a_{ii}^* = b_{ii} \Leftrightarrow \Gamma(\epsilon_{i0}^* \circ \widehat{a_{ii}} \circ \epsilon_{1j}^*) \perp \Gamma(-\epsilon_{1j} \circ \widehat{b_{ii}} \circ \epsilon_{i0}).$$

Proposition 4.5.11. *The map*

$$\rho = \eta \circ \theta : R \to End(V, \langle \cdot, \cdot \rangle)$$

*defined in Corollary 4.5.3 is a *-ring-embedding of involutive rings with unit.*

▶ **Proof.** We recall that the map

$$\eta : R(\Phi, L, M) \hookrightarrow End(V_D)$$

defined in Corollary 4.4.7 was defined via the lattice embedding $\iota : L \hookrightarrow L(V_D)$. In this situation, we consider a MOL L and a MOL-embedding of L into $L(V_D, \langle ., . \rangle)$. As shown before, for morphisms $\epsilon_{ji} : M_i \to M_j \in \Phi$, we can define an adjoint operator $\epsilon_{ji}^* : M_j \to M_i$ via the orthogonality on L. For morphisms $\varphi_{10} : M_0 \to M_1$ and $\psi_{01} : M_1 \to M_0$, the relation of adjointness can be captured via orthogonality on graphs. In particular, we have $\rho(\epsilon_{ji}^*) = \big(\rho(\epsilon_{ji})\big)^*$ and $\rho(\varphi_{10}^*) = \big(\rho(\varphi_{10})\big)^*$.

Now, for $r \in R$, consider $e_j r e_i \in e_j R e_i$. Then

$$
\begin{aligned}
\big(\rho(e_j r e_i)\big)^* &= \big(\rho(e_{j1}(e_{1j}e_j r e_i e_{i0})e_{0i})\big)^* \\
&= \big(\rho(e_{j1})\rho(e_{1j}e_j r e_i e_{i0})\rho(e_{0i})\big)^* \\
&= \big(\rho(e_{0i})\big)^*\big(\rho(e_{1j}e_j r e_i e_{i0})\big)^*\big(\rho(e_{j1})\big)^* \\
&= \rho(e_{0i}^*)\rho\big((e_{1j}e_j r e_i e_{i0})^*\big)\rho(e_{j1}^*) \\
&= \rho(e_{0i}^*)\rho\big(e_{i0}^* e_i^* r^* e_j^* e_{1j}^*\big)\rho(e_{j1}^*) \\
&= \rho\big(e_{0i}^* e_{i0}^* e_i^* r^* e_j^* e_{1j}^* e_{j1}^*\big) = \rho(e_i^* r^* e_j^*) \\
&= \rho\big((e_j r e_i)^*\big).
\end{aligned}
$$

Hence, we have

$$
\begin{aligned}
\left(\rho(r)\right)^* &= \left(\rho(\sum e_j r e_i)\right)^* = \left(\sum \rho(e_j r e_i)\right)^* \\
&= \sum \left(\rho(e_j r e_i)\right)^* = \sum \rho((e_j r e_i)^*) \\
&= \sum \rho(e_i^* r^* e_j^*) = \rho\left(\sum e_i^* r^* e_j^*\right) \\
&= \rho\left(\sum (e_j r e_i)^*\right) = \rho\left(\left(\sum e_j r e_i\right)^*\right) = \rho(r^*).
\end{aligned}
$$

◄

Simple *-Regular Rings

Corollary 4.5.12. *Each simple *-regular ring S with unit is faithfully representable.*

▷ Proof. We may assume that S is non-Artinian, hence, we can assume that S has height at least 3. By Corollary 1.5.6, the MOL $L = \overline{L}(S_S)$ contains a stable orthogonal frame of format (n, k) with $n \geq 3$. It follows by Proposition 4.5.11 that S is faithfully representable. ◁

Proposition 4.5.13. *Every simple *-regular ring has a faithful linear positive representation.*

▶ **Proof.** The case of a simple *-regular ring with unit was handled in Corollary 4.5.12. So, let us consider a simple *-regular ring R without unit.

Consider the set $P(R)$ of all projections in R. Since R is *-regular, $P(R)$ is a lattice. In particular, we can consider $P(R)$ as a directed set. By Lemma 1.5.11, we have that R is the directed union of its subrings R_e, $e \in P(R)$. By Lemma 4.2.8, R is a *-regular subring of an ultraproduct of the R_e, $e \in P(R)$. By Lemma 1.5.10, for each projection e, the ring R_e is a simple ∗-regular ring with unit e. By Corollary 4.5.12, each R_e admits a faithful linear positive representation. Hence, we can conclude with Lemma 4.2.9 and Proposition 4.2.10 that R has a faithful linear positive representation. ◄

Representations of Ideals

Observation 4.5.2. Assume that I is a two-sided ideal of a ∗-regular ring R. Recalling Lemma 4.2.1, we can consider I as a ∗-regular ring on its own (without unit, if I is non-trivial). Thus, we can consider representations of I.

Proposition 4.5.14. *Let R be a ∗-regular ring, I a two-sided ideal in R and $\varrho : I \to End(V_D, \langle \cdot, \cdot \rangle)$ a representation of I. Denote the set of all projections in I by $P(I)$, abbreviate $V_p := \varrho(p)[V]$ for a projection $p \in P(I)$ and set*

$$\rho(r) := \bigcup_{p \in P(I)} \varrho(rp)_{|V_p}$$

Then ρ is a representation of R an appropriate subspace U of V, where the scalar product on U_D is given by restriction.

▶ **Proof.** Firstly, we have to show that the given definition of ρ indeed defines a map $\rho : R \to End(V_D)$. Recalling that the set of all projections of a ∗-regular ring is directed, we consider two projections $e, f \in P(I)$ with $e \leq f$, that is, with $e = fe$. We have to show that the restrictions coincide on V_e, i.e., $\varrho(rf)_{|V_e} = \varrho(re)_{|V_e}$.

Since $e = fe$, we have

$$\varrho(re)_{|V_e} = \varrho(rfe)_{|V_e} = (\varrho(rf) \circ \varrho(e))_{|V_e} = \varrho(rf)_{|V_e},$$

as desired.

Secondly, we have to show that the map

$$\rho : R \to End(V_D) \qquad r \mapsto \bigcup_{p \in P(I)} \varrho(rp)_{|V_p}$$

is a ∗-ring homomorphism. For 0 in R, we have

$$\rho(0) = \bigcup_{p \in P(I)} \varrho(0p)_{|V_p} = 0_V.$$

If $1 \in R$, we have

$$\rho(1) = \bigcup_{p \in P(I)} \varrho(1p)_{|V_p} = 1_U$$

with $U := \bigcup_{p \in P(I)} V_p$, that is, $\rho[R]$ acts on the subspace U of V_D.

For addition, let $r, s \in R$. We have

$$
\begin{aligned}
\rho(r + s) &= \bigcup_{p \in P(I)} \varrho((r + s)p)_{|V_p} = \bigcup_{p \in P(I)} \varrho(rp + sp)_{|V_p} \\
&= \bigcup_{p \in P(I)} \varrho(rp)_{|V_p} + \varrho(sp)_{|V_p} \\
&= \bigcup_{p \in P(I)} \varrho(rp)_{|V_p} + \bigcup_{q \in P(I)} \varrho(sq)_{|V_q} \\
&= \rho(r) + \rho(s).
\end{aligned}
$$

Furthermore, we can conclude that $\rho(-r) = -\rho(r)$.

For multiplication, let $r, s \in R$. We note that for each $p \in P(I)$, there exists $q_p \in P(I)$ such that $sp = q_p sp$. We claim that the following holds.

$$\bigcup_{p \in P(I)} (\varrho(rq_p) \circ \varrho(sp))_{|V_p} = \bigcup_{q \in P(I)} \varrho(rq)_{|V_q} \circ \bigcup_{p \in P(I)} \varrho(sp)_{|V_p}$$

▷ **Proof.** Take $v \in U$. Then there exists $p_v \in P(I)$ with $v \in V_{p_v}$. Thus, on one hand, we have

$$\left(\bigcup_{p \in P(I)} (\varrho(rq_p) \circ \varrho(sp))_{|V_p} \right)(v) = (\varrho(rq_{p_v}) \circ \varrho(sp_v))(v),$$

while on the other hand

$$\left(\bigcup_{q \in P(I)} \varrho(rq)_{|V_q} \circ \bigcup_{p \in P(I)} \varrho(sp)_{|V_p} \right)(v)$$

$$= \bigcup_{q \in P(I)} \varrho(rq)_{|V_q} \left(\bigcup_{p \in P(I)} \varrho(sp)_{|V_p}(v) \right)$$

$$= \bigcup_{q \in P(I)} \varrho(rq)_{|V_q} \left(\varrho(sp_v)(v) \right)$$

$$= \varrho(rq_{p_v})(\varrho(sp_v)(v)).$$

◁

Hence,

$$
\begin{aligned}
\rho(rs) &= \bigcup_{p \in P(I)} \varrho((rs)p)_{|V_p} = \bigcup_{p \in P(I)} \varrho(rq_p sp)_{|V_p} \\
&= \bigcup_{p \in P(I)} (\varrho(rq_p) \circ \varrho(sp))_{|V_p} \\
&= \bigcup_{q \in P(I)} \varrho(rq)_{|V_q} \circ \bigcup_{p \in P(I)} \varrho(sp)_{|V_p} \\
&= \rho(r) \circ \rho(s).
\end{aligned}
$$

Now we examine the involution on R. For $r \in R$, consider $v, w \in U$. Then take $e \in P(I)$ with $v \in V_e$. There exists $f_1 \in P(I)$ such that $f_1 re = re$ and $f_2 \in P(I)$ such that $w \in V_{f_2}$. Choosing $f := f_1 \vee f_2$, we have

$$
\begin{aligned}
\langle \rho(r)v, w \rangle &= \langle \bigcup_{p \in P(I)} \varrho(rp)_{|V_p} v, w \rangle = \langle \varrho(re)v, w \rangle \\
&= \langle \varrho(fre)v, w \rangle = \langle v, \varrho(er^* f)w \rangle \\
&= \langle v, \varrho(e)\varrho(r^* f)w \rangle = \langle \varrho(e)v, \varrho(r^* f)w \rangle \\
&= \langle v, \varrho(r^* f)w \rangle = \langle v, \rho(r^*)w \rangle,
\end{aligned}
$$

where we have used that $\varrho : I \to End(V_D)$ is a ∗-ring homomorphism and $v \in V_e, w \in V_f$. ◄

Lemma 4.5.15. *Let R be a ∗-regular ring and I a two-sided ideal in R. Assume that I has a linear positive representation $\varrho : I \to End(V_D, \langle ., . \rangle)$. Denote the action of R on the ideal I given by left multiplication by λ_I, that is*

$$
\lambda_I : R \to End(I_I) \qquad \lambda_I(r)(x) := \widehat{r}(x) = rx
$$

If the representation $\varrho : I \rightarrow End(V_D, \langle \cdot, \cdot \rangle)$ is faithful and $\lambda_I : R \rightarrow End(I_I)$ is injective, then the representation $\rho : R \rightarrow End(V_D, \langle \cdot, \cdot \rangle)$ defined in Proposition 4.5.14 is faithful.

▷ **Proof.** Assume that $\rho(r) = 0$. This is equivalent to $\varrho(re) = 0$ for all $e \in P(I)$. As $\varrho : I \rightarrow End(V_D, \langle \cdot, \cdot \rangle)$ is faithful, this means that $re = 0$ for all $e \in P(I)$. Since I is a *-regular ring, for every element $x \in I$ there exists $e \in P(I)$ such that $ex = x$. Hence, we have that $rx = 0$ for all $x \in I$. Since we assumed the action of R on I given by left multiplication to be injective, we have that $r = 0$. This shows that ρ is injective. ◁

Subdirectly Irreducible *-Regular Rings

In this section, we will show that each subdirectly irreducible *-regular ring R has a faithful representation.

Observation 4.5.3. We may assume that R is non-Artinian: Since every regular ring is semiprime, a subdirectly irreducible *-regular ring which is Artinian is semisimple, hence representable.

Furthermore, the minimal two-sided ideal of R is non-Artinian, too (see [HS, Proposition 2]).

Proposition 4.5.16. *Let R be a subdirectly irreducible *-regular ring with minimal two-sided ideal J.*

Then the action $\lambda_J : R \rightarrow End(J_J)$ of R on J defined by

$$\lambda_J(r)(x) = \hat{r}(x) = rx$$

is injective.

▶ **Proof.** Consider the left annihilator $A := ann^l_R(J)$ of J in R. We claim that A is closed under left and right multiplication by elements of R.

▷ Proof. Let $a \in A, r \in R, b \in J$. On the one hand, we have

$$(ra)(b) = r(ab) = r0 = 0,$$

since $ab \in J$. Therefore, $ra \in A$. On the other hand, we have

$$(ar)b = a(rb) = 0,$$

since $rb \in J$. Therefore, $ar \in A$. ◁

Since A is closed under addition, it follows that A is a two-sided ideal in R. We claim that J does not annihilate itself.

▷ Proof. Take $a \in J$, $a \neq 0$. Consider the principal right ideal aR generated by a in R. Then $aR = pR$ for a projection $p \in R$. With $pR = aR$, we have $p = ax$ for some $x \in R$, so $p \in J$. Hence $p = p^2 \in J^2$, so $J^2 \neq \{0\}$. ◁

Now, since J is the minimal two-sided ideal in R and J does not annihilate itself, we can conclude that A is trivial. Therefore, the action of R on J defined by left multiplication is injective. ◀

Lemma 4.5.17. *The minimal ideal J of a subdirectly irreducible *-regular ring R is a simple *-regular ring.*

▷ Proof. Let A be an ideal in J, $A \neq \{0\}$. Consider the two-sided ideal $B := \langle A \rangle_R$ generated by A in R. Since $\{0\} \neq A \subseteq B \subseteq J$ and J is minimal, we have that $B = J$. Since $J^2 = J$, it follows that

$$J = JJJ = JBJ \subseteq A \subseteq J.$$

Thus, $A = J.$ ◁

Remark 4.5.8. Note that we did not need to assume that R contains a unit – even without a unit element, we can sandwich A, as the following shows.

Consider $x \in B = \langle A \rangle_R$ and $b, c \in J$, so $bxc \in JBJ$. But $x \in B$ means that

$$x = a + \sum r_i a_i + \sum a_j s_j + \sum r_k a_k s_k$$

for some $r_i, s_j, r_k, s_k \in R$ and $a, a_i, a_j, a_k \in A$, so

$$bxc = bac + \sum (br_i)(a_i c) + \sum (ba_j)(s_j c) + \sum (br_k) a_k (s_k c).$$

But then

$$br_i, s_j c, br_k, s_k c \in J \quad \text{and} \quad bac, a_i c, ba_j \in A,$$

hence $bxc \in A$.

For $J^2 = J$, consider $a \in J$. Since R is ∗-regular, there exists a projection $p \in R$ with $pR = aR$. That is, there exists $y \in R$ such that $p = ay$. In particular, $p \in J$, so $pa = a \in J^2$, i.e., $J \subseteq J^2$. The converse inclusion is trivial.

Lemma 4.5.18. *Let R be a ∗-regular ring and I a minimal two-sided ideal in R. Let e be a projection in the minimal ideal I (which in particular is simple as a ring).*
Then the ring $R_e = eRe$ is simple.

▷ Proof. Let A be a two-sided ideal in R_e. Since $e \in I$, we have $R_e \subseteq I$, thus $A \subseteq I$. Consider the two-sided ideal $\langle A \rangle_I$ generated by A in I. If $\langle A \rangle_I$ does not vanish, then $\langle A \rangle_I = I$. In particular, every element $x \in eRe \subseteq I$ belongs to $\langle A \rangle_I$. Consequently, we have $A = eRe$. ◁

Proposition 4.5.19. *Considered as simple *-regular ring, the minimal ideal J of a subdirectly irreducible *-regular ring R has a faithful representation.*

▷ Proof. Proposition 4.5.13. ◁

Theorem 4.5.20. *Every subdirectly irreducible *-regular ring R has a faithful representation.*

▷ Proof. Lemma 4.5.15. ◁

4.6 Varieties of *-Regular Rings

In this section, we want to prove that *every* variety of *-regular rings is generated by its simple Artinian members. We begin with some preliminary definitions and results.

Preliminaries

The following can be found in [Jón60, Section 2] or [Mic03].

Definition 4.6.1. Arguesian lattice
A lattice L is called *Arguesian* if the following condition is satisfied for arbitrary elements $a_0, a_1, a_2, b_0, b_1, b_2$ in L. For

$$y := (a_0 + a_1)(b_0 + b_1)\big[(a_0 + a_2)(b_0 + b_2) + (a_1 + a_2)(b_1 + b_2)\big],$$

we have

$$(a_0 + b_0)(a_1 + b_1)(a_2 + b_2) \le a_0(a_1 + y) + b_0(b_1 + y).$$

Remark 4.6.1. Jónsson points out that the name was suggested because a projective geometry is Desarguean if and only if its subspace lattice is Arguesian.

Definition 4.6.2. ∗-Coordinatisation of a lattice

A regular ring R is said to *coordinatise* the CML L if its lattice of principal right ideals $\overline{L}(R_R)$ is lattice-isomorphic to L.

A ∗-regular ring R is said to ∗-*coordinatise* the MOL L if $\overline{L}(R_R)$ is MOL-isomorphic to L, where the ortho-complement in $\overline{L}(R_R)$ is given as in Section 1.5.

We call a CML (a MOL) L *(∗-)coordinatisable* if there exists a (∗-)regular ring R (∗-)coordinatising L.

Remark 4.6.2. Of course, this notions of coordinatisability makes sense for sectionally complemented modular lattices and sectionally modular ortholattices, when the coordinatising rings are non-unital.

Remark 4.6.3. Every lattice of commuting equivalence relations is Arguesian. Every coordinatisable lattice is Arguesian [Jón60].

The following results can be found in [Mic03].

Corollary 4.6.3. Coordinatisation of MOLs

Every Arguesian simple MOL of height at least three is ∗-coordinatisable by a simple ∗-regular ring with unit.

Especially, every simple MOL of height at least four is ∗-coordinatisable by a simple ∗-regular ring with unit.

Theorem 4.6.4. Varieties of MOLs

Every variety of Arguesian MOLs is generated by simple members L_k of height $h(L_k) \leq 2$ and members of the form $\overline{L}(R_R)$ for a simple ∗-regular ring R.

▷ Proof. See [Mic03, Theorem 4.17] and [HR99, Corollary 4.6]. ◁

Definition 4.6.5. 2-distributive lattice
A lattice L is called *2-distributive* if it satisfies the following equation.

$$x_0 \cdot (x_1 + x_2 + x_3) = x_0(x_2 + x_3) + x_0(x_1 + x_3) + x_0(x_1 + x_2)$$

For the more general notion of n-distributivity of a lattice, see for example [HMR05].

Remark 4.6.4. Note that the property of a lattice to be *Arguesian* is defined by an inequality, and the property to be *2-distributive* is defined by an equality. Therefore, both properties are preserved under homomorphic images, substructures and products (the class operators H, S and P).

Reduction I

As before, we consider countable, subdirectly irreducible, non-Artinian, *-regular rings.

Proposition 4.6.6. Approximation
*Let R be a countable, subdirectly irreducible, non-Artinian, *-regular ring. Then*

$$R \in V\big(\{S_k : k \in K | S_k \in V(R),$$
$$S_k \text{ simple with unit of height } h(S_k) \geq 3\}\big).$$

Remark 4.6.5. The line of arguments will shed light on the choice of the name of Proposition 4.6.6.

▶ **Proof.** We pick up the thread of Section 4.5 and Proposition 4.5.14 and recall the following. The minimal two-sided ideal J of R is a ∗-regular ring on its own. We have already shown that J is simple non-Artinian. Hence, we consider J as a simple, non-Artinian, ∗-regular ring without unit.

We will make use of Lemma 4.5.18: For each projection e in J, the subring $R_e = eRe$ is a simple ∗-regular ring with unit e. Since J is non-Artinian, each subring R_e is contained in another subring R_f with $f \in P(J)$ of height at least 3. In particular, the MOL $\overline{L}(R_{f R_f})$ of all principal right ideals of the simple ∗-regular ring R_f contains a stable orthogonal (n, k)-frame with $n \geq 3$.

Since R is a subdirectly irreducible ∗-regular ring, R has a faithful linear positive representation in a vector space V. As a subring of R, the two-sided ideal J admits a faithful linear positive representation in V, too. As in Proposition 4.5.14, J gives rise to a representation space U of R, where U is a subspace of V. Hence, R can be considered as a subring of $End(U)$.

Now, we follow the approach of Tyukavkin and Micol (compare [Mic03, Section 3.2]), but instead of using the *externally given* matrices p_n (where p_n projects onto $span(e_1, \ldots e_n)$), we use the *internally given* projections $e \in P(J)$. Recall that, on the one hand, this set of projections is countable, and on the other hand, the representation space U of R is given by $U := \bigcup_{e \in P(J)} V_e$, where $V_e := \rho(e)[V]$. Hence, we use the set $P(J)$ of all projections in J to construct an ascending countable chain $\mathfrak{C} := \{p_i : i \in I\}$, where $p_i \in P(J)$ and $I \leq \mathbb{N}$, such that this chain exhausts the space U.

For an index $i \in I$ and a projection $p_i \in \mathfrak{C}$, we define the ring $R_i := p_i R p_i$. We set $S := \prod_{i \in I} R_i$. As usual, we define the ring operations on the ring S componentwise.

We say that a net $(a_i) \in S$ converges to an element $a \in R$ if the following condition is satisfied.

$$\forall l \in I. \exists n \in I. \forall m \geq n. \qquad a_m p_l = a p_l \wedge p_l a_m = p_l a$$

Remark 4.6.6. This notion of convergence is chosen analogously to the notion given in [Mic03, Section 3.2], the only difference being the index set I instead of the set of all natural numbers.

Lemma 4.6.7. *If a net $(a_i)_{i \in I} \in S$ converges to $a \in R$, then this limit is unique.*

▷ **Proof.** Assume that $(a_i)_{i \in I} \in S$ converges to both a and b in R. Let $l \in I$ be given. Then there exists $n \in I$ such that for all $m \geq n$, we have $a_m p_l = a p_l$ and $p_l a_m = p_l a$. Thus, for each $l \in I$, we have $a p_l = b p_l$. We recall that $U := \bigcup_{e \in P(J)} V_e$, and that in particular the projections $p_i \in \mathfrak{C}$ exhaust the space U. Consequently, a and b act identically on U. Hence, $a = b$. ◁

Lemma 4.6.8. *For each a in R, there exists a net $(a_i)_{i \in I}$ in S that converges to a.*

▷ **Proof.** Let $a \in R$. As in [Mic03, Proposition 3.14 (2)], we define the desired net $(a_i)_{i \in I}$ via $a_i := p_i a p_i$, where $p_i \in \mathfrak{C}$. Let $l \in I$ be given. Since J is a two-sided ideal, we have $a p_l \in J$. Since J is a *-regular ring, there exists $p_k \in \mathfrak{C}$ such that $(a p_l) p_k = a p_l$ and $p_k (p_l a) = p_l a$. In particular, for $n := max(k, l)$, we have $(p_n a p_n) p_l = a p_l$ and $p_l (p_n a p_n) = p_l a$, yielding the desired index $n \in I$. Therefore, the net $(a_i)_{i \in I}$ converges to a. ◁

Lemma 4.6.9. *Taking limits is compatible with the ring operations* $+, -, \cdot, ^{*}$.

▷ Proof. The proof is similar to the proof of [Mic03, Prop. 3.14 (3)]. Just take the index set I instead of $\mathbb{N}$. ◁

Hence, we have shown that the ring R is contained in $HSP\big(\{e_i Re_i : e_i \in P(J)\}\big)$. Since $e_i Re_i$ is a subring of R, trivially $e_i Re_i \in HSP(R)$. By Lemma 4.5.18, each subring $e_i Re_i$ with $e_i \in P(J)$ is simple. As pointed out at the beginning of the section, each subring $e_i Re_i$ is contained in a subring $e_k Re_k$ of height at least 3. Consequently, the MOL of all principal right ideals of $e_k Re_k$ contains a stable orthogonal (n, k)-frame with $n \geq 3$. ◀

Reduction II

Proposition 4.6.10. *Let L be a simple MOL of height at least four or of height at least three and Arguesian. Then*

$$V(L) = V\big(\{L_k : k \in K | L_k \in V(L),$$
$$L_k \text{ simple Artinian with } h(L_k) \geq 3\}\big).$$

For the proof, we rely on Section 4.6 and [HR99].

▶ **Proof.** By Corollary 4.6.3, a simple MOL L of height at least four (or height at least three and Arguesian) is $*$-coordinatisable by a simple $*$-regular ring R. In any case, L is Arguesian. By [HR99], L lies in the MOL-variety generated by its Artinian members. Clearly, we can reduce to simple Artinian members L_k, since an Artinian MOL decomposes into simple Artinian ones. Furthermore, each L_k is Arguesian since it lies in the variety generated by the Arguesian MOL L.

Now we consider the heights of the MOLs L_k. We know that the MOLs of height one or two are embeddable into MOLs of height three. Hence, the only thing left to show is that $V(L)$ is not generated by MOLs L_k of height less than or equal to two only.

We recall that a MOL of height less than or equal to two is 2-distributive. Thus, if all L_k would have height one or two only, every member of the generated variety would be 2-distributive, too. But this is not possible, since the MOL L has $h(L) \geq 3$ and thus is not 2-distributive. ◀

Generation of a Variety of *-Regular Rings

Lemma 4.6.11. *Let S and T be *-regular rings. Assume that $\overline{L}(S_S)$ contains a stable orthogonal (n, k)-frame with $n \geq 3$ and*

$$\varphi : \overline{L}(S_S) \to [0, u] \leq \overline{L}(T_T)$$

is a morphism of MOLs.

*Then there exists a morphism $\widehat{\varphi} : S \to T$ of *-regular rings such that φ is induced by $\widehat{\varphi}$ via $\varphi(eS) = \widehat{\varphi}(e)T$.*

In particular, $ker(\varphi)$ corresponds to $ker(\widehat{\varphi})$ in the following way: $ker(\varphi) = \{eS : \widehat{\varphi}(e) = 0\}$.

▷ Proof.   Due to Lemma 1.5.9, we have $u = eT$ and $[0, u] \cong \overline{L}(T_{eT_e})$. Hence we can assume without loss of generality that $\varphi : \overline{L}(S_S) \to \overline{L}(T_T)$, where T is a *-regular ring. Moreover, the forward image $\varphi[\Phi]$ of the frame Φ under φ is a stable orthogonal frame in $\overline{L}(T_T)$ of the same format as Φ.

By Section 4.4, Proposition 4.4.6 and Corollary 4.4.7 and Section 4.5, Corollary 4.5.2, we get a morphism of rings $\widehat{\varphi} : S \to T$. Since we can capture the involution of the rings S and T via the induced orthogonality relations on $\overline{L}(S_S)$ and $\overline{L}(T_T)$ and since $\varphi : \overline{L}(S_S) \to \overline{L}(T_T)$ is a morphism of MOLs, we can use Section 4.5 and the generalised result of Proposition 4.5.11 to conclude that $\widehat{\varphi} : S \to T$ is a $*$-ring morphism. $\triangleleft$

Corollary 4.6.12. *Let S and S_k with $k \in K$ be $*$-regular rings. Assume that $\overline{L}(S_S)$ contains a stable orthogonal (n, k)-frame Φ with $n \geq 3$, and that $\overline{L}(S_S)$ is contained in the variety $V\big(\{\overline{L}(S_{k\,S_k}) : k \in K\}\big)$.*

Then S is contained in $V\big(\{S_k : k \in K\}\big)$.

▷ **Proof.** $\overline{L}(S_S)$ is the homomorphic image of a sublattice of the product $\prod \overline{L}(S_{k\,S_k})$:

$$\overline{L}(S_S) \twoheadleftarrow M \hookrightarrow \prod_{k \in K} \overline{L}(S_{k\,S_k})$$

Since stable orthogonal frames of a fixed format are projective (see Section 1.4, Lemma 1.4.18), we can conclude that there exists a MOL $M' \leq M$ containing a stable orthogonal (n, k)-frame Ψ of the same format as Φ such that the restriction of the morphism $\gamma : M \to \overline{L}(S_S)$ maps Ψ onto Φ.

Since the MOL $M' \leq M$ is a sublattice of the coordinatised lattice $\prod \overline{L}(S_{k\,S_k})$ and since Ψ is a stable orthogonal (n, k)-frame in M' with $n \geq 3$, we can conclude that M' is coordinatisable, i.e., $M' \cong \overline{L}(T_T)$, where T is a *-regular ring.

Identifying $\overline{L}(T_T)$ with M' and M with its image in $\prod \overline{L}(S_{kS_k})$, we consider the restrictions

$$\pi_{k|_{M'}} : \overline{L}(T_T) \to \overline{L}(S_{kS_k})$$

of the projections π_k. Since the forward image $\pi_{k|_{M'}}[\Psi]$ of the stable orthogonal frame Ψ is either a stable orthogonal frame of the same format or is collapsed to a point, we can use Lemma 4.6.11 to get morphism $\widehat{\pi_k} : T \to S_k$ of *-regular rings, where k runs over an index set $K' \subseteq K$ such that $\pi_k[\Psi]$ is non-trivial for all $k \in K'$.

Using Lemma 4.6.11 again, we can conclude that

$$\bigcap_{k \in K'} ker(\widehat{\pi_k}) = \bigcap_{k \in K'} ker(\pi_k) = \{0\},$$

that is, $T \in SP(\{S_k : k \in K'\})$. Consequently,

$$S \in H(T) = HSP(\{S_k : k \in K'\}),$$

where we have used Lemma 4.6.11 a final time. ◁

Theorem 4.6.13. *Let V be a variety of *-regular rings. Then*

$$V = V(\{S_k : k \in K | S_k \in V, \quad S_k \text{ simple Artinian }\}).$$

▶ **Proof.** As in Section 4.6, it is enough to consider $V = V(R)$, where R is a countable subdirectly irreducible *-regular ring. Since a *-regular ring R is a directed union of its *-regular subrings eRe, $e \in P(R)$, we can assume that R is a ring with unit.

Due to Lemma 1.3.15, every subdirectly Artinian ∗-regular ring is simple. Hence, we can reduce to the case that R is non-Artinian.

By Proposition 4.6.6, we can reduce to the case that R is a non-Artinian simple ring with unit and $\overline{L}(R_R)$ contains a stable orthogonal (n, k)-frame with $n \geq 3$. Since $\overline{L}(R_R)$ is Arguesian, Proposition 4.6.10 yields

$$\overline{L}(R_R) \;\in\; V\big(\{L_k : k \in K | L_k \in V(\overline{L}(R_R)),$$
$$L_k \text{ simple Artinian with } h(L_k) \geq 3\}\big).$$

As in the proof of Proposition 4.6.10, each L_k is Arguesian. Hence, by Corollary 4.6.3, L_k is ∗-coordinatisable, i.e., $L_k \cong \overline{L}(S_{k S_k})$ for some ∗-regular ring S_k. Because $\overline{L}(S_{k S_k})$ is a simple MOL with $h(L_k) \geq 3$, L_k contains a stable orthogonal (n, k)-frame with $n \geq 3$. By Corollary 4.6.12, $S_k \in V(R)$. Since $\overline{L}(R_R)$ contains a stable orthogonal (n, k)-frame with $n \geq 3$, Corollary 4.6.12 yields that

$$R \in V\big(\{S_k : k \in K | S_k \in V. \in V, \quad S_k \text{ simple Artinian }\}\big).$$

◀

5 On Atomic Rings

If you want to understand a thing,
then take it apart.
Reduce it to its components.

— Unknown

In this chapter, we follow the approach of [Mic03, Chapter 3] and consider the structure of atomic regular involutive rings. We begin with the examination of the ideal structure of regular involutive rings, which is more complicated than in the case of *-regular rings. In particular, we have to deal with pathological cases.

5.1 Structure Theory

In this section, we want to examine the ideal structure of regular involutive rings. We will point out some differences between regular involutive rings and *-regular rings, in particular, we will stretch the difference between (two-sided) ideals and (two-sided) *-ideals.

Ideals and ∗-Ideals

Let R be a ring. A two-sided ideal of R is a subset of R which is closed under addition and under left and right multiplication with elements of R. If R is equipped with an involution, we define the notion of a two-sided ∗-ideal.

Definition 5.1.1. Two-sided ∗-ideal
Let R be an involutive ring. We call a subset I of R a *two-sided ∗-ideal* if it is a two-sided ideal and if it is closed under the involution $* : R \to R$.

If R is ∗-regular, every two-sided ideal is a two-sided ∗-ideal. For general regular involutive rings, this is not the case, as the following example shows.

Definition 5.1.2. Exchange-involutive ring
Let S be a ring and T be a ring which is anti-isomorphic to S.[1] Let $f : S \to T$ and $g : T \to S$ be the corresponding anti-isomorphisms (i.e., f, g are mutually inverse to each other). Consider the direct product $R := S \times T$ of S and T and and define an involution on R via

$$* : R \to R \qquad (a, b) \mapsto (g(b), f(a))$$

We call this involution the *exchange involution* on R with respect to f and g and R the *full exchange-involutive ring* with respect to S, T and f, g.

Any involution of similar form (e.g. with the maps f and g not necessarily specified) will be called an *exchange-involution*.

We call a ring R' an *exchange-involutive ring* if R' is a subdirect product of S and T and the involution on

---

[1] E.g., T could be the opposite ring S^{op}.

R' is given by an exchange involution on $S \times T$, i.e., if R' is $*$-embeddable into a full exchange involutive ring $R = S \times T$ and R' is a subdirect product of the factors S and T.

Subdirect and $*$-Subdirect Irreducibility

In this section, we will examine the notion of subdirect irreducibility in the context of regular involutive rings and $*$-regular rings. We define the notion of a subdirectly irreducible algebra **A** in the usual way.

Definition 5.1.3. Subdirect irreducibility
An algebra **A** is called *subdirectly irreducible* if for every subdirect embedding

$$\alpha : \mathbf{A} \to \prod_{i \in I} \mathbf{A}_i$$

there exists an $i \in I$ such that

$$\pi_i \circ \alpha : \mathbf{A} \to \mathbf{A}_i$$

is an isomorphism.

Observation 5.1.1. In the context of involutive rings, we have to distinguish between subdirect irreducibility and $*$-subdirect irreducibility, because the concept of subdirectly irreducibility depends on the considered algebraic structure. Since every involutive ring R can be considered as an algebra $(R, +, \cdot, ^*, 0, 1)$ (where the involution is a unary operation) or as an algebra $(R, +, \cdot, 0, 1)$, we have to distinguish between the properties being *sub-*

directly irreducible as a ring and being *subdirectly irreducible as an involutive ring.*

Following the characterisation of subdirect irreducibility given by [UA00, Theorem 8.4], and excluding the pathological case (of a trivial ring), we have the following.

Corollary 5.1.4. *An involutive ring is subdirectly irreducible iff it has a smallest two-sided ideal. An involutive ring is ∗-subdirectly irreducible iff it has a smallest two-sided ∗-ideal.*

Observation 5.1.2. Let R be an involutive ring. Then R can be ∗-subdirectly irreducible without being subdirectly irreducible. For example, let R be an exchange-involutive ring where each factor is a simple ring. Then the only two-sided ∗-ideal is the whole ring, but each factor is a two-sided ideal on its own - but of course not a two-sided ∗-ideal.

In [Mic03] Chapter 3, many results have been proved for (atomic) subdirectly irreducible ∗-regular rings. This section is devoted to the hope that some of them will also hold for (atomic) subdirectly irreducible regular involutive rings.

To follow the approach of [Mic03], we have to decompose a regular involutive ring into its subdirectly irreducible factors. Hence, we have to face the difficulty of deciding between the following options: Either we decompose R into a subdirect product of subdirectly irreducible factors (which might not be involutive rings) or we decompose R into a subdirect product of ∗-subdirectly irreducible factors (which are involutive rings on their own, but might not be subdirectly irreducible rings).

We think the following approach is the most promising: We decompose an atomic regular involutive ring R into factors which are involutive rings on their own right, with an involution inherited from R. We are guided by the hope that the following holds:

Either such a factor is subdirectly irreducible as a ring (and we can derive results similar to the ones in [Mic03]) or the factor is isomorphic to an exchange-involutive ring.

Exchange Rings

Lemma 5.1.5. *Let R be an involutive ring and A be a minimal $*$-ideal which is not a minimal ideal.*

Then A decomposes into a direct sum $A = I + I^*$, where $I \subseteq A$ is a non-trivial ideal in R.

▷ Proof. Since A is not a minimal ideal, there exists an ideal $I \subseteq A$ such that $I \neq \{0\}$ and $I \neq A$. Since A is a $*$-ideal, the image I^* of I under the involution is a subset of A as well. The sum $I + I^*$ is a subset of A which is indeed a two-sided $*$-ideal of R. Since $I + I^*$ is non-trivial, we can conclude that $A = I + I^*$. Since the intersection $I \cap I^*$ is a $*$-ideal which is a subset of A, it has to be trivial, that is, the sum of I and I^* is direct. ◁

Lemma 5.1.6. *Let R be a regular involutive ring. Assume that R is a subdirect product of two anti-isomorphic factors such that the involution on R is given by an exchange involution.*

Then each factor is a regular ring on its own.

▷ **Proof.** Immediate, since the factors S and T are homomorphic images of R. ◁

Lemma 5.1.7. *Let R be an involutive ring. Assume that R is $*$-subdirectly irreducible but not subdirectly irreducible.*

Then R is isomorphic to an exchange-involutive ring with subdirectly irreducible factors.

▷ **Proof.** Since R is $*$-subdirectly irreducible but not subdirectly irreducible, there exists a minimal two-sided $*$-ideal A which is not a minimal ideal. As shown above, A is a direct sum of a two-sided ideal I and its involutive image I^*.

We consider the product $S := R/I \times R/I^*$ and define an involution on S via

$$ * : S \to S : \qquad (r/I, s/I^*) \mapsto (s^*/I, r^*/I^*). $$

We consider the map

$$ f : R \to S \qquad r \mapsto (r/I, r/I^*). $$

Since $I \cap I^* = \{0\}$, this map is injective; of course, it is a ring homomorphism. Furthermore, the map

$$ g : R/I \to R/I^* \qquad r/I \mapsto r^*/I^* $$

is an anti-isomorphism. Therefore, R is embedded into an exchange-involution ring. ◁

Atomic Reg. Inv. Rings and Representability

Definition 5.1.8. Atomic regular ring

Let R be a regular ring and $\overline{L}(R_R)$ its lattice of principal right ideals. If $\overline{L}(R_R)$ is an atomic lattice, we call R an *atomic* regular ring.

Remark 5.1.1. A more precise term would be *right* atomic. But since we consider only the lattice of principal right ideals, we drop the prefix *right* for simplicity. In that, we follow [Mic03]. As in [Mic03, Chapter 3], we note that the property of a ring to be *atomic* can be expressed in first-order logic.

In [Mic03], Micol proves the following result.

Proposition 5.1.9. *Let R be a $*$-regular ring and $q \in \overline{L}(R_R)$ an atom.*
Then there exists a two-sided ($*$-)ideal J_q in R such that $J_q \cap q = \{0\}$ and R/J_q is a $*$-subdirectly irreducible atomic $*$-regular ring.

In the case of regular involutive rings, the analogon is

Lemma 5.1.10. *Let R be a regular involutive ring and $q \in \overline{L}(R_R)$ an atom.*
Then there exists a two-sided $*$-ideal J_q in R such that $J_q \cap q = \{0\}$ and R/J_q is a $*$-subdirectly irreducible atomic regular involutive ring.
Furthermore, the lattice homomorphism induced by the projection $\pi_q : R \to R/J_q$ maps q onto an atom.

▷ Proof. Let $a \neq 0$ be an element in R. Then there exists a two-sided $*$-ideal J_a which is maximal with respect to the property $a \notin J_a$ (J_a corresponds to a $*$-ring congruence maximal with $a \not\equiv 0$). Hence, the factor ring R/J_a is $*$-subdirectly irreducible.

Now, let $q \in \overline{L}(R_R)$ be an atom. Choose an element a in R with $q = aR$ and set $J_q := J_a$ and $R_q := R/J_q$. Then $J_q \cap q = \{0\}$.

As in [Mic03, Lemma 3.1], we can use [Mic03, Theorem 1.4] to conclude that the map $\Pi : \overline{L}(R_R) \to \overline{L}(R_{q R_q})$ induced by $\pi : R \to R_q$ maps q to an atom. In particular, R_q has at least one atom. Hence, with [HR99, Proposition 2.17], the lattice $\overline{L}(R_{q R_q})$ is atomic. Thus, the ring R_q is atomic, as desired. ◁

Corollary 5.1.11. *In the situation above, the factor $R_q = R/J_q$ is either subdirectly irreducible as a ring or it is an exchange-involutive ring.*

Micol then proves the following result [Mic03, Theorem 3.2].

Proposition 5.1.12. *Let R be an atomic $*$-regular ring.*

Then R is a ($*$-)subdirect product of ($*$-)subdirectly irreducible $*$-regular atomic rings.

In the case of regular involutive rings, the analogon is

Proposition 5.1.13. *Let R be an atomic regular involutive ring.*

Then R is a subdirect product of atomic $*$-subdirectly irreducible regular involutive rings, where each factor is either an atomic regular involutive ring (which is subdirectly irreducible) or an atomic regular exchange-involutive ring (which is $*$-subdirect irreducible, but not subdirectly irreducible).

▶ **Proof.** Let $q \in \overline{L}(R_R)$ be an atom. Then there exists a two-sided $*$-ideal J_q such that $J_q \cap q = \{0\}$ and R/J_q

is a $*$-subdirectly irreducible atomic regular involutive ring. By Corollary 5.1.11, R/J_q is either a subdirectly irreducible atomic regular involutive ring or an exchange-involutive ring.

Let us denote the set of all atoms in $\overline{L}(R_R)$ by $A(R)$. Since the intersection $\mathcal{J}(R) := \bigcap_{q \in A(R)} J_q$ is a right ideal containing no atom and R is atomic, we have that $\mathcal{J}(R) = \{0\}$. Hence, we get a subdirect ring-embedding

$$\alpha : R \to \prod_{q \in A(R)} R_q$$

compatible with the involution. ◄

[Mic03, Corollary 3.4] reads as follows.

Corollary 5.1.14. *A subdirectly irreducible atomic $*$-regular ring is representable, that is, it admits a faithful linear positive representation.*

For this context, the appropriate analogon is

Lemma 5.1.15. *A subdirectly irreducible atomic regular ring R is representable as a ring, that is, it admits a faithful linear ring representation.*

▷ Proof. Following [Mic03], we rely on Jacobson: Due to [Jac89], Chapter 4, a subdirectly irreducible atomic regular ring is primitive. Hence, R has a faithful linear representation. ◁

Definition 5.1.16. Weak g-representability
An involutive ring is called *weakly g-representable* if it is isomorphic to a subring of a direct product of linear

representable rings R_i and rings S_k, where each S_k is
an exchange-involutive ring $S_k = S_k^1 \times S_k^2$ and S_k^i is a
subring of an endomorphism ring of a vector space.

Theorem 5.1.17. *Every atomic regular involutive ring
is weakly g-representable.*

▶ **Proof.** We combine the previous results. An atomic
regular involutive ring decomposes into a ∗-subdirect product of ∗-subdirectly irreducible atomic regular involutive
rings by Proposition 5.1.13. If an atomic ∗-subdirectly irreducible factor is subdirectly irreducible, it has a faithful linear representation. If an atomic ∗-subdirectly irreducible factor is not subdirectly irreducible, it is isomorphic to an exchange-involutive ring. Consequently, R is
weakly g-representable. ◀

6 Appendix

Fields of characteristic two should be forbidden.

— Anonymous

This chapter is dedicated to different subjects. In the first section, we will present an alternative and elementary approach to questions of representability of regular involutive rings. The point of concern is the flawed proof given by Herstein for [Hst76, Theorem 1.2.1] that caused some trouble. Our technique takes far enough to handle everything but the case of skew fields of characteristic two.

In the second section, we will mention an alternative approach for Theorems 2.1.2 and 2.1.3.

6.1 An Alternative Approach

This section falls into line with Chapter 3.

We will use Proposition 3.1.1 as a starting point, aiming at the following.

Proposition 6.1.1. *Let R be a primitive involutive ring with minimal left ideal $A = Re$ with an idempotent gen-*

erator $e \in R$. Assume that R does not contain minimal idempotent Hermitian elements.

Then the following holds: Either the ideal A has the property

$$\forall x \in A. \qquad x^* x = 0$$

or the ring $S := (e + e^*)R(e + e^*)$ is isomorphic to a matrix ring $M(2, D)$ over a field D with $char(D) = 2$, where $D \cong eRe$ and the involution on D induced by the involution on R is trivial and the involution on $M(2, D)$ is given by

$$* : M(2, D) \to M(2, D) \qquad \begin{pmatrix} a & b \\ c & d \end{pmatrix} \mapsto \begin{pmatrix} d & b \\ c & a \end{pmatrix}.$$

Proof of Proposition 6.1.1

In this section, we will present a proof of Proposition 6.1.1, divided into multiple subsections. Firstly, we will use the minimal idempotent to introduce a 2×2-matrix ring over a skew field. Secondly, we will determine the restriction of the involution of R to this matrix ring. Thirdly, we will distinguish the cases of $char(D) \neq 2$ and $char(D) = 2$.

Introduction of the Matrix Ring

Proposition 6.1.2. *Let R be a primitive involutive ring. Assume that*

1. *R contains no minimal projections,*

2. *R contains an idempotent, non-Hermitian element e which generates a minimal left ideal Re.*

Due to the result of Herstein (see Chapter 3, Proposition 3.1.1), we can assume that $ee^* = 0$.

Then the following holds: If we set

$$f := e - e^*e + e^*,$$

we obtain an idempotent Hermitian element $f \in R$ such that $S := fRf$ is isomorphic to a matrix ring $Mat(2, D)$ over some field D. If furthermore $char(D) \neq 2$, we have $x^*x = 0$ for all $x \in Re$.

▶ **Proof.** Since e is idempotent, so is e^*. Furthermore, we have

$$
\begin{aligned}
f^2 &= (e - e^*e + e^*)^2 \\
&= e^2 - ee^*e + ee^* \\
&\quad - e^*ee - e^*ee^*e \\
&\quad - e^*ee^* + e^*e - e^*e^*e + e^*e^* \\
&= e + 0 + 0 - e^*e - 0 - 0 + e^*e - e^*e + e^* \\
&= e - e^*e + e^* = f
\end{aligned}
$$

and $f^* = f$, i.e., f is a projection. We define

$$g := -e^*e + e$$

and notice that

$$g^2 = e^*ee^*e - e^*ee + e(-e^*e) + ee = 0 - e^*e - 0 + e = g.$$

Furthermore, $e^*g = 0 = ge^*$. Considering the elements e, f and g, we see that $fe = e = ef$ and $fg = g = gf$. Thus, e and g are elements of the ring $S := fRf$. Since f

is Hermitian, we have that e^*, g^* are elements of S, too. Therefore, we get a decomposition of $1_{fRf} = f$ into a sum of commuting idempotent elements, namely $f = g + e^*$. Then we can conclude with Lemma 1.2.8 that the ring $S := fRf$ is isomorphic to a matrix ring $Mat(2, D)$ over some skew field D. ◄

Observation 6.1.1. Up to now, we have no further information concerning D or the action of the involution of R restricted to S. Nevertheless, with Proposition 3.1.1, we can show the following lemma.

Lemma 6.1.3. *Let $X \in Mat(2, D) \neq 0$ be singular. Then $XX^* = 0$ or $X^*X = 0$.*

▷ **Proof.** If $X \in M(2, D)$ is singular, then $\psi^{-1}(X)$ generates a minimal one-sided ideal. Hence, we have that $\psi^{-1}(X) \cdot (\psi^{-1}(X))^* = 0$ or $(\psi^{-1}(X))^* \cdot \psi^{-1}(X) = 0$. Consequently, $XX^* = 0$ or $X^*X = 0$. ◁

The Involution on $Mat(2, D)$

In the following section, we want to determine the action of the involution on R on the ring S – or rather on its isomorphic image $M(2, D)$.

We define a map $\psi : S \to Mat(2, D)$ in the following way. We define

$$e^* \mapsto A = \begin{pmatrix} 1 & 0 \\ 0 & 0 \end{pmatrix} \quad \text{and} \quad g \mapsto B = \begin{pmatrix} 0 & 0 \\ 0 & 1 \end{pmatrix}.$$

Setting

$$X^* := \psi((\psi^{-1}(X))^*),$$

we obtain an involution on $Mat(2, D)$.

We want to determine this involution on $Mat(2, D)$ exactly. To do so, we will first examine the involution on the matrices A and B (corresponding to the summands e^* and g in $1_S = e^* + g$).

The Matrix A^* We start with the general form

$$A^* = \begin{pmatrix} a & b \\ c & d \end{pmatrix}.$$

Since $ee^* = 0$, we have $A^*A = 0$, i.e.,

$$A^*A = \begin{pmatrix} a & b \\ c & d \end{pmatrix} \cdot \begin{pmatrix} 1 & 0 \\ 0 & 0 \end{pmatrix} = \begin{pmatrix} a & 0 \\ c & 0 \end{pmatrix} = 0.$$

Therefore, $a = c = 0$ and

$$A^* = \begin{pmatrix} 0 & b \\ 0 & d \end{pmatrix}.$$

Since e and e^* are idempotent, so are A^* and A. We calculate

$$(A^*)^2 = \begin{pmatrix} 0 & bd \\ 0 & d^2 \end{pmatrix},$$

hence

$$A^* = \begin{pmatrix} 0 & b \\ 0 & d \end{pmatrix} = \begin{pmatrix} 0 & bd \\ 0 & d^2 \end{pmatrix}.$$

Because $A^* \neq 0$, we can conclude that $d \neq 0$. Since D is a skew field, the equality $d^2 = d$ yields $d = 1$. We have arrived at

$$A^* = \begin{pmatrix} 0 & b \\ 0 & 1 \end{pmatrix}.$$

The Matrix AA^* Since e^*e is Hermitian, so is

$$AA^* = \begin{pmatrix} 1 & 0 \\ 0 & 0 \end{pmatrix} \cdot \begin{pmatrix} 0 & b \\ 0 & 1 \end{pmatrix} = \begin{pmatrix} 0 & b \\ 0 & 0 \end{pmatrix},$$

i.e.,

$$\begin{pmatrix} 0 & b \\ 0 & 0 \end{pmatrix}^* = \begin{pmatrix} 0 & b \\ 0 & 0 \end{pmatrix}.$$

Furthermore, we have

$$g^* = (-e^*e + e)^* = -e^*e + e^*,$$

so

$$\begin{pmatrix} 0 & 0 \\ 0 & 1 \end{pmatrix}^* = -AA^* + A = \begin{pmatrix} 1 & -b \\ 0 & 0 \end{pmatrix}.$$

Possible Values for b We assume that the adjoint of $A = \begin{pmatrix} 0 & 0 \\ 0 & 1 \end{pmatrix}$ is given by

$$\begin{pmatrix} 0 & 0 \\ 0 & 1 \end{pmatrix}^* = \begin{pmatrix} 1 & -b \\ 0 & 0 \end{pmatrix}$$

with $b \neq 0$.

In the following, we want to calculate the adjoint of

$$X := \begin{pmatrix} 0 & 0 \\ b^{-1} & 0 \end{pmatrix}$$

in $M(2, D)$. Again, we start with the general form

$$\begin{pmatrix} 0 & 0 \\ b^{-1} & 0 \end{pmatrix}^* = \begin{pmatrix} x & y \\ u & v \end{pmatrix}.$$

We have

$$\begin{pmatrix} 0 & b \\ 0 & 0 \end{pmatrix} \cdot \begin{pmatrix} 0 & 0 \\ b^{-1} & 0 \end{pmatrix} = \begin{pmatrix} 1 & 0 \\ 0 & 0 \end{pmatrix}.$$

Starring both sides, we get

$$\begin{pmatrix} 0 & 0 \\ b^{-1} & 0 \end{pmatrix}^* \cdot \begin{pmatrix} 0 & b \\ 0 & 0 \end{pmatrix}^* = \begin{pmatrix} 1 & 0 \\ 0 & 0 \end{pmatrix}^*.$$

Using the already derived results

$$\begin{pmatrix} 0 & b \\ 0 & 0 \end{pmatrix}^* = \begin{pmatrix} 0 & b \\ 0 & 0 \end{pmatrix} \qquad \begin{pmatrix} 1 & 0 \\ 0 & 0 \end{pmatrix}^* = \begin{pmatrix} 0 & b \\ 0 & 1 \end{pmatrix}$$

and plugging in

$$\begin{pmatrix} 0 & 0 \\ b^{-1} & 0 \end{pmatrix}^* = \begin{pmatrix} x & y \\ u & v \end{pmatrix},$$

we get

$$\begin{pmatrix} x & y \\ u & v \end{pmatrix} \cdot \begin{pmatrix} 0 & b \\ 0 & 0 \end{pmatrix} = \begin{pmatrix} 0 & b \\ 0 & 1 \end{pmatrix}.$$

Calculating the product on the left hand side, we get

$$\begin{pmatrix} 0 & xb \\ 0 & ub \end{pmatrix} = \begin{pmatrix} 0 & b \\ 0 & 1 \end{pmatrix}.$$

Therefore, $x = 1, u = b^{-1}$.

Now, we consider the equality

$$\begin{pmatrix} 0 & 0 \\ 0 & 1 \end{pmatrix} \cdot \begin{pmatrix} 0 & 0 \\ b^{-1} & 0 \end{pmatrix} = \begin{pmatrix} 0 & 0 \\ b^{-1} & 0 \end{pmatrix}.$$

Starring, we get

$$\begin{pmatrix} 0 & 0 \\ b^{-1} & 0 \end{pmatrix}^{*} \cdot \begin{pmatrix} 0 & 0 \\ 0 & 1 \end{pmatrix}^{*} = \begin{pmatrix} 0 & 0 \\ b^{-1} & 0 \end{pmatrix}^{*}.$$

Therefore,

$$\begin{pmatrix} 1 & y \\ b^{-1} & v \end{pmatrix} \cdot \begin{pmatrix} 1 & -b \\ 0 & 0 \end{pmatrix} = \begin{pmatrix} 1 & y \\ b^{-1} & v \end{pmatrix}.$$

Calculating the product on the left hand side, we get

$$\begin{pmatrix} 1 & -b \\ b^{-1} & -1 \end{pmatrix} = \begin{pmatrix} 1 & y \\ b^{-1} & v \end{pmatrix}$$

and so $y = -b, v = -1$, i.e., we have arrived at

$$\begin{pmatrix} 0 & 0 \\ b^{-1} & 0 \end{pmatrix}^{*} = \begin{pmatrix} 1 & -b \\ b^{-1} & -1 \end{pmatrix}.$$

Since

$$X = \begin{pmatrix} 0 & 0 \\ b^{-1} & 0 \end{pmatrix}$$

is singular, we can conclude that $XX^{*} = 0$ or $X^{*}X = 0$ by Lemma 6.1.3. We calculate both products and get

$$XX^* = \begin{pmatrix} 0 & 0 \\ b^{-1} & 0 \end{pmatrix} \cdot \begin{pmatrix} 0 & 0 \\ b^{-1} & 0 \end{pmatrix}^*$$

$$= \begin{pmatrix} 0 & 0 \\ b^{-1} & 0 \end{pmatrix} \cdot \begin{pmatrix} 1 & -b \\ b^{-1} & 1 \end{pmatrix} = \begin{pmatrix} 0 & 0 \\ b^{-1} & -1 \end{pmatrix} \neq 0$$

and

$$X^*X = \begin{pmatrix} 1 & -b \\ b^{-1} & 1 \end{pmatrix} \cdot \begin{pmatrix} 0 & 0 \\ b^{-1} & 0 \end{pmatrix} = \begin{pmatrix} -1 & 0 \\ b^{-1} & 0 \end{pmatrix} \neq 0.$$

Therefore, the assumption $b \neq 0$ leads to a contradiction. Hence, we have shown that the only possible value for $b \in D$ is $b = 0$. Consequently, for $A = diag(1,0)$, we have $A^* = diag(0,1)$. In particular, we have $A^* = B$ and $A^*A = 0$.

More Adjoints   Using the previous results, we want to determine the involution on S. We consider

$$X = \begin{pmatrix} a & 0 \\ 0 & 0 \end{pmatrix} \in Mat(2, D).$$

This matrix satisfies

$$X \cdot \begin{pmatrix} 1 & 0 \\ 0 & 0 \end{pmatrix} = X = \begin{pmatrix} 1 & 0 \\ 0 & 0 \end{pmatrix} \cdot X.$$

Starring, we get

$$\begin{pmatrix} 0 & 0 \\ 0 & 1 \end{pmatrix} \cdot X^* = X^* = X^* \cdot \begin{pmatrix} 0 & 0 \\ 0 & 1 \end{pmatrix}.$$

Again, starting with

$$X^* = \begin{pmatrix} x & y \\ u & v \end{pmatrix},$$

we get

$$\begin{pmatrix} 0 & 0 \\ 0 & 1 \end{pmatrix} \cdot X^* = \begin{pmatrix} 0 & 0 \\ u & v \end{pmatrix}$$

and

$$X^* \cdot \begin{pmatrix} 0 & 0 \\ 0 & 1 \end{pmatrix} = \begin{pmatrix} 0 & y \\ 0 & v \end{pmatrix},$$

so we see that

$$X^* = \begin{pmatrix} 0 & 0 \\ 0 & \alpha_{11}(a) \end{pmatrix},$$

where $a_{11} : D \to D$ is some map.

Considering the matrix

$$Y = \begin{pmatrix} ab & 0 \\ 0 & 0 \end{pmatrix} \in Mat(2, D),$$

we get

$$(\begin{pmatrix} ab & 0 \\ 0 & 0 \end{pmatrix})^* = \begin{pmatrix} 0 & 0 \\ 0 & \alpha_{11}(ab) \end{pmatrix}$$

$$= [\begin{pmatrix} a & 0 \\ 0 & 0 \end{pmatrix} \cdot \begin{pmatrix} b & 0 \\ 0 & 0 \end{pmatrix}]^* = \begin{pmatrix} b & 0 \\ 0 & 0 \end{pmatrix}^* \begin{pmatrix} a & 0 \\ 0 & 0 \end{pmatrix}^*$$

$$= \begin{pmatrix} 0 & 0 \\ 0 & \alpha_{11}(b) \end{pmatrix} \begin{pmatrix} 0 & 0 \\ 0 & \alpha_{11}(a) \end{pmatrix} = \begin{pmatrix} 0 & 0 \\ 0 & \alpha_{11}(b)\alpha_{11}(a) \end{pmatrix}.$$

Therefore, $\alpha_{11} : D \to D$ is an anti-automorphism. Analogously, one can show that

$$\begin{pmatrix} 0 & 0 \\ 0 & d \end{pmatrix}^* = \begin{pmatrix} a_{22}(d) & 0 \\ 0 & 0 \end{pmatrix},$$

where $\alpha_{22} : D \to D$ is another anti-automorphism on D. Since $* : S \to S$ is an involution, we have the relations

$$\alpha_{22} \circ \alpha_{11} = id_D = \alpha_{11} \circ \alpha_{22}.$$

For the non-diagonal entries, we obtain

$$\begin{pmatrix} 0 & b \\ 0 & 0 \end{pmatrix}^* = \begin{pmatrix} 0 & \alpha_{12}(b) \\ 0 & 0 \end{pmatrix} \qquad \begin{pmatrix} 0 & 0 \\ c & 0 \end{pmatrix}^* = \begin{pmatrix} 0 & 0 \\ \alpha_{21}(c) & 0 \end{pmatrix},$$

where $\alpha_{12}, \alpha_{21} : D \to D$ are additive maps such that $\alpha_{12}^2 = \alpha_{21}^2 = id_D$. Summarising, we have

$$\begin{pmatrix} a & b \\ c & d \end{pmatrix}^* = \begin{pmatrix} \alpha_{11}(d) & \alpha_{12}(b) \\ \alpha_{21}(c) & \alpha_{22}(a) \end{pmatrix},$$

where α_{11}, α_{22} are anti-automorphisms and α_{12}, α_{21} additive maps on D satisfying

$$\alpha_{12}^2 = \alpha_{21}^2 = id_D \qquad \text{and} \qquad \alpha_{22} \circ \alpha_{11} = id_D.$$

The Maps α_{ik}   We prove the following relations.

1. $\alpha_{12}(bd) = \alpha_{22}(d) \circ \alpha_{12}(b)$

2. $\alpha_{12}(1) \cdot \alpha_{21}(1) = 1 = \alpha_{21}(1) \cdot \alpha_{12}(1)$

3. $\alpha_{12}(1) = -1$

4. $\alpha_{12} = -\alpha_{22}$

$\alpha_{12}(bd) = \alpha_{22}(d) \circ \alpha_{12}(b)$.   We consider

$$\begin{pmatrix} 0 & b \\ 0 & 0 \end{pmatrix} \begin{pmatrix} 0 & 0 \\ 0 & d \end{pmatrix} = \begin{pmatrix} 0 & bd \\ 0 & 0 \end{pmatrix}.$$

Starring, we get

$$\begin{pmatrix} \alpha_{22}(d) & 0 \\ 0 & 0 \end{pmatrix} \begin{pmatrix} 0 & \alpha_{12}(b) \\ 0 & 0 \end{pmatrix} = \begin{pmatrix} 0 & \alpha_{12}(bd) \\ 0 & 0 \end{pmatrix}.$$

$\alpha_{12}(1) \cdot \alpha_{21}(1) = 1 = \alpha_{21}(1) \cdot \alpha_{12}(1)$.   We consider

$$\begin{pmatrix} 0 & 1 \\ 1 & 0 \end{pmatrix}^2 = \begin{pmatrix} 1 & 0 \\ 0 & 1 \end{pmatrix}.$$

Starring and using

$$\begin{pmatrix} 0 & 1 \\ 1 & 0 \end{pmatrix}^* = \begin{pmatrix} 0 & \alpha_{12}(1) \\ \alpha_{21}(1) & 0 \end{pmatrix} \quad \text{and} \quad \begin{pmatrix} 1 & 0 \\ 0 & 1 \end{pmatrix}^* = \begin{pmatrix} 1 & 0 \\ 0 & 1 \end{pmatrix},$$

we get

$$\begin{pmatrix} 0 & \alpha_{12}(1) \\ \alpha_{21}(1) & 0 \end{pmatrix}^2 = \begin{pmatrix} 1 & 0 \\ 0 & 1 \end{pmatrix}.$$

Therefore,

$$\alpha_{12}(1) \cdot \alpha_{21}(1) = 1 = \alpha_{21}(1) \cdot \alpha_{12}(1).$$

$\alpha_{12}(1) = -1$.   We consider

$$X := \begin{pmatrix} \alpha_{12}(1) & \alpha_{12}(1) \\ 1 & 1 \end{pmatrix}.$$

Starring and using the fact that α_{22} is an anti-automorphism, so $\alpha_{22}(1) = 1$ and the fact that $\alpha_{12}^2 = id_D$, we get

$$X^* = \begin{pmatrix} 1 & 1 \\ \alpha_{21}(1) & \alpha_{11}(\alpha_{12}(1)) \end{pmatrix}.$$

Since the second row is a multiple of the first, X is singular, hence, $XX^* = 0$ or $X^*X = 0$ by Lemma 6.1.3. We determine both products and get

$$XX^* = \begin{pmatrix} \alpha_{12}(1) + \alpha_{12}(1) \cdot \alpha_{21}(1) & * \\ * & * \end{pmatrix}$$

and

$$X^*X = \begin{pmatrix} \alpha_{12}(1) + \alpha_{12}(1) \cdot \alpha_{21}(1) & * \\ * & * \end{pmatrix}.$$

Using the result above, we can conclude that

$$XX^* = \begin{pmatrix} \alpha_{12}(1) + 1 & * \\ * & * \end{pmatrix} \quad \text{and} \quad X^*X = \begin{pmatrix} \alpha_{12}(1) + 1 & * \\ * & * \end{pmatrix},$$

so we arrive at $\alpha_{12}(1) = -1$.

$\alpha_{12} = -\alpha_{22}$. We consider $\alpha_{12}(bd) = \alpha_{22}(d) \cdot \alpha_{12}(b)$. Choosing $b = 1$ and using $\alpha_{12}(1) = -1$, we arrive at $\alpha_{12}(d) = -\alpha_{22}(d)$.

Summary Calculating similar relations, we get the following:

$$\begin{pmatrix} a & b \\ c & d \end{pmatrix}^* = \begin{pmatrix} \alpha(d) & -\alpha(b) \\ -\alpha(c) & \alpha(a) \end{pmatrix},$$

where $\alpha : D \to D$ is an anti-automorphism of order 2.

$\alpha = id$, D **is commutative**   We consider

$$X = \begin{pmatrix} 1 & a \\ a & a^2 \end{pmatrix}.$$

Since the second row is a multiple of the first, X is singular. Therefore, we can conclude that $XX^* = 0$ or $X^*X = 0$. We calculate both products and get

$$X^*X = \begin{pmatrix} \alpha(a^2) & -\alpha(a) \\ -\alpha(a) & 1 \end{pmatrix} \cdot \begin{pmatrix} 1 & a \\ a & a^2 \end{pmatrix} = \begin{pmatrix} * & * \\ -\alpha(a) + a & * \end{pmatrix}$$

and

$$XX^* = \begin{pmatrix} 1 & a \\ a & a^2 \end{pmatrix} \cdot \begin{pmatrix} \alpha(a^2) & -\alpha(a) \\ -\alpha(a) & 1 \end{pmatrix} = \begin{pmatrix} * & -\alpha(a) + a \\ * & * \end{pmatrix}.$$

Thus $\alpha(a) - a = 0$, i.e., $\alpha = id_D$. Therefore,

$$\begin{pmatrix} a & b \\ c & d \end{pmatrix}^* = \begin{pmatrix} d & -b \\ -c & a \end{pmatrix}$$

and D is commutative.

   Now, having shown that D is a field, we can introduce a determinant $\det : S \to D$ and use it to show the following.

Lemma 6.1.4. *Let $B \in S$ be singular.*
   Then $BB^* = 0$ and $B^*B = 0$.

▷ **Proof.** Assume that

$$B = \begin{pmatrix} a & b \\ c & d \end{pmatrix} \in S$$

is singular. Since we have introduced a determinant, singularity of B is equivalent to $det(B) = 0$. We calculate

$$BB^* = \begin{pmatrix} a & b \\ c & d \end{pmatrix} \cdot \begin{pmatrix} d & -b \\ -c & a \end{pmatrix} = \begin{pmatrix} ad - bc & -ab + ba \\ cd - dc & -cb + da \end{pmatrix}.$$

Using commutativity of the field, we get

$$BB^* = diag(ad{-}bc, -bc{+}ad) = diag(det(B), det(B)) = 0.$$

Analogously, we can show that $B^*B = 0$. ◁

Remark 6.1.1. Note that, before the introduction of a determinant, we just had the implication

$$X \text{ singular} \implies XX^* = 0 \text{ or } X^*X = 0.$$

Conclusion for $char(D) \neq 2$

Approaching the end of the proof of Proposition 6.1.1, we distinguish the cases of $char(D) = 2$ and $char(D) \neq 2$. In this section, assume $char(D) \neq 2$.

Consider an arbitrary element $x \in S$ such that $x^* = x$, i.e.,

$$\begin{pmatrix} a & b \\ c & d \end{pmatrix} = \begin{pmatrix} d & -b \\ -c & a \end{pmatrix}.$$

Since $char(D) \neq 2$, the equations $b = -b$ and $c = -c$ imply that $b = 0 = c$. Furthermore, we have $a = d$, so $x = diag(a, a)$. But since scalar multiples of the identity lie in the centre of $M(2, S)$, we have that

$$AxA^* = xAA^* = 0$$

and
$$A^* x A = x A^* A = 0,$$

where $A = diag(1,0)$ is the matrix from the very beginning. That is, if $char(D) \neq 2$ and $x \in S$ is Hermitian, we have
$$A x A^* = 0 = A^* x A.$$

Now consider $r \in R$ arbitrary. Using
$$fe = e \quad e^* f^* = e^* \quad f^* = f,$$

we get
$$(re)^*(re) = e^* r^* re = e^* f^* r^* rfe = e^*(fr^* rf)e = 0,$$

since $fr^* rf$ lies in S and is Hermitian. Therefore,
$$\forall r \in R. \qquad (re)^*(re) = 0.$$

Conclusion for $char(D) = 2$

Again, we consider an arbitrary element $x \in S$ such that $x^* = x$, i.e.,
$$\begin{pmatrix} a & b \\ c & d \end{pmatrix} = \begin{pmatrix} d & -b \\ -c & a \end{pmatrix}.$$

In this case, we cannot conclude that x is a scalar multiple of the identity matrix. The above equation only yields that $d = a$; since $char(D) = 2$, we can omit the factor -1. Therefore, we have
$$x = \begin{pmatrix} a & b \\ c & a \end{pmatrix}.$$

Calculating, we get

$$
\begin{aligned}
AxA^* &= \begin{pmatrix} 1 & 0 \\ 0 & 0 \end{pmatrix} \cdot \begin{pmatrix} a & b \\ c & a \end{pmatrix} \cdot \begin{pmatrix} 0 & 0 \\ 0 & 1 \end{pmatrix} \\
&= \begin{pmatrix} 1 & 0 \\ 0 & 0 \end{pmatrix} \cdot \begin{pmatrix} 0 & b \\ 0 & a \end{pmatrix} = \begin{pmatrix} b & 0 \\ 0 & 0 \end{pmatrix} \neq 0.
\end{aligned}
$$

Therefore, the line of argument used in the previous section *cannot* be used to conclude that for a Hermitian element in S the relation $e^*xe = 0$ holds. In particular, we cannot use the previous reasoning to conclude that for arbitrary $r \in R$, the element $re \in A$ has the desired property that $(re)^*(re) = 0$.

This finishes the proof of Proposition 6.1.1.

6.2 Countable Representable Rings

In Chapter 2, Section 2.1, we have shown that a countable regular involutive with a faithful linear representation admits a faithful linear representation in a vector space of at most countable dimension, by presenting a construction method for the representation space.

Another possibility to show the desired result is the following. Let $\sigma = (D, V_D, \phi, \rho)$ be the representation of the countable regular involutive ring R. We consider (R, V_D) as a multi-sorted structure, where R is a regular involutive ring and V_D a right vector space over the involutive skew field D, equipped with the non-degenerated form ϕ. Then we extend the signature of (R, V_D) to incorporate the relation that R admits a faithful linear representation in V_D.

That is, for each ring element $r \in R$ we integrate the linear map $\rho(r) : V \to V$ into the signature of (R, V_D), extending it by countable many constants. Furthermore, we express the conditions that ϕ is a non-degenerated (symplectic or $*$-Hermitian) form on V_D, that the map $\rho : R \to End(V_D)$ is a morphism of involutive rings, and that the action of R on V_D is injective.

Since R is countable, the extended signature is still countable. Hence, there exists a countable model; that is, R admits a faithful linear representation in a vector space of at most countable dimension.

7 Open Questions

*We are at the very beginning
of time for the human race.
It is not unreasonable
that we grapple with problems.
But there are tens of thousands of years in the future.
Our responsibility is to do what we can,
learn what we can, improve the solutions,
and pass them on.*

— Richard Feynman

So, where to continue? Now that we have proven that every $*$-regular ring is representable in an appropriate sense, that we have shown that every variety of $*$-regular rings is generated by its simple Artinian members, that we have completed the characterisation of representability of regular involutive rings, and that we have carried on the analysis of atomic regular involutive rings and their decompositions into subdirectly irreducible components, one might ask: What is left to do? But of course, there is a multitude of questions about regular involutive and $*$-regular rings.

One question is concerned with the finer circumstances of representations of ∗-regular rings. Even though we know that every ∗-regular ring R is a subdirect product of (subdirectly irreducible ∗-regular rings and hence) representable ∗-regular rings, we do not know yet the possibilities of R to be embedded into such a product.

Another topic would the finiteness (see Chapter 2) of regular involutive and ∗-regular rings. In this thesis, some results have been mentioned (see the Prologue), others sketched or touched (see Chapter 2, Section 2.4). At the beginning of this research, there was only the faintest hope that the approach via representability and representations would lead to substantial improvements in this area. In retrospective, it is still unknown whether the results and the techniques of this thesis might be of any help for questions of (direct) finiteness of regular involutive and ∗-regular rings. In this context, one should consult the textbook [Good91] and in particular, the article [CL90], where Chuang and Lee construct a certain regular ring R. On the one hand, R is a subdirect product of simple Artinian rings; more exactly, R is constructed as a subring of a product $W = S \times \prod_{k \in \mathbb{N}} E_k$, where S is a subring of the full endomorphism ring $End(V_F)$ of a vector space of countable dimension over a field F, and E_k the set of $k \times k$-block matrices. (Similar to Chapter 2, E_k is defined via $E_k := \pi_k End(V_F)\pi_k$, where p_k is the projection of V onto the span of the first k basis vectors of a certain basis of V.) In particular, the ring R is faithfully g-representable in our sense. But on the other hand, R is not unit-regular. Maybe this example can inspire further research about regular involutive rings, representability and unit-regularity.

Bibliography

Books and Collected Works

[Berb72] S. K. Berberian, *Baer ∗-Rings* Springer, 1972.

[UA00] S. N. Burris, H. P. Sankappanavar, *A Course in Universal Algebra: The Millenium Edition* http://www.math.uwaterloo.ca/~snburris.

[Cohn01] P. M. Cohn, *An Introduction to Ring Theory* Springer, 2nd printing, 2001.

[Good91] K. R. Goodearl, *Von Neumann Regular Rings* Krieger Publishing Company, 1991.

[Gor98] V. A. Gorbunov, *Algebraic Theory of Quasivarieties* Siberian School of Logic and Algebra, Consultants Bureau, New York, 1998.

[Gross79] H. Gross, *Quadratic Forms in Infinite Dimensional Vector Spaces* Birkhäuser, Boston, 1979.

[Jac64] N. Jacobson, *Structure of Rings* American Mathematical Society, 2nd edition, 1964.

[Jac85] N. Jacobson, *Basic Algebra I* W. H. Freeman and Company, 2nd edition, 1985.

[Jac89] N. Jacobson, *Basic Algebra II* W. H. Freeman and Company, 2nd edition, 1989.

215

[Hst76]   I. N. Herstein, *Rings with Involution* The University of Chicago Press, Chicago, 1976.

[Kap68]   I. Kaplansky, *Rings of Operators* W. A. Benjamin, Inc., 1968.

[Lam01]   T. Y. Lam *A First Course in Non-Commutative Rings* Springer, 2nd edition, 2001.

[Mae58]   F. Maeda *Kontinuierliche Geometrien* Springer, Berlin, 1958.

[Row88]   L. Rowen, *Ring Theory, Volume I* Academic Press, Inc., 1988.

[Skor64]   L. Skornyakov, *Complemented Modular Lattices and Regular Rings* Oliver and Boyd Ltd., 1964.

[vdWII]   B. L. van der Waerden, *Moderne Algebra II* Springer, Berlin, 1931.

[vN60]   J. von Neumann, *Continuous Geometries* Princeton, 1960.

[MN]   F. J. Murray, J. von Neumann, *On Rings of Operators* Annals of Mathematics, Transactions of the American Mathematical Society, 1936–1949.

Articles and Theses

[AM84]   P. Ara, P. Menal, *On regular rings with involution* Arch. Math., 42, 1984, p. 126–130.

[Ber57]   S. K. Berberian, *The regular ring of a finite AW*-algebra* Ann. of Math., 65, 1957, p. 224–240.

[Ber72]   S. K. Berberian, *The regular ring of a finite Baer-*-ring* Journal of Algebra, 23, 1972, p. 35–65.

[CL90]   C.-L. Chuang, P.-H. Lee, *On regular subdirect products of simple Artinian rings* Pacific Journal of Mathematics, 142 (1), 1990, p. 17–21.

[Day82]   A. Day, *Geometrical applications in modular* Universal Algebra and Lattice Theory (R. Freese and O. Garcia, eds.). In: Lecture Notes in Mathematics, 1004, 1983, p. 111–141, Springer.

[Day84]   A. Day, *Applications of coordinatization in modular lattice theory: The legacy of J. von Neumann* 1, 1984, p. 295–300, Reidel Publishing Company, Boston, U.S.A.

[FH64]   L. Fuchs, I. Halperin, *On the imbedding of a regular ring in a regular ring with identity* Fundamenta Mathematicae, 54, 1964, p. 285–290.

[Good82]   K. R. Goodearl, *Directly finite $\aleph_0$-continuous regular rings* Pacific Journal of Mathematics, 100 (1), 1982, p. 105–122.

[GMM93]   K. R. Goodearl, P. Menal, J. Moncasi, *Free and residually Artinian regular rings* Journal of Algebra, 156 (2), 1993, p. 407–432.

[Hand77]   D. Handelman, *Completions of rank rings* Canad. Math. Bull., 20 (2), 1977, p. 199–205.

[Hol70]   S. S. Holland, Jr., *The current interest in orthomodular lattices* Trends in Lattice Theory, 1970, p. 41–126, Van Nostrand Reinhold Company.

[Herr95]   C. Herrmann, *Alan Day's work on modular and Arguesian lattices* Algebra Universalis, 34, 1995, p. 35–60. Dedicated to the memory of Alan Day.

[Herr]   C. Herrmann, *Representation and atomic extension of modular ortholattices*   to appear.

[HMR05]   C. Herrmann, F. Micol, M. Roddy *On n-distributive modular ortholattices* Algebra Universalis, 53, 2005, p. 143–147. Dedicated to the memory of Ivan Rival.

[HR99]   C. Herrmann, M. Roddy, *Proatomic modular ortholattices: Representation and equational theory* Note di matematica e fisica, 10, 1999, p. 57–88.

[HS]   C. Herrmann, M. Semenova, *Existence varieties of regular rings and complemented modular lattices*   to appear.

[Jón60]   B. Jónsson, *Representations of complemented modular lattices* Transactions of the American Mathematical Society, 97, 1960, p. 64–94.

[Kap55]   I. Kaplansky, *Any orthocomplemented complete modular lattice is a continuous geometry* Annals of Mathematics, 61, 1955, p. 524–542.

[Mic03]   F. Micol, *On representability of *-regular rings and modular ortholattices* FB 04 TUD, 2003.

[Nie03]   N. Niemann, *Representations of involutive rings* FB 04 TUD, 2003.

[vN36]   J. von Neumann *On regular rings* Proc. Natl. Acad. Science U.S.A., 22, 1936, p. 707–713.

[vN58]   J. von Neumann *The non-isomorphism of certain continuous rings* Annals of Mathematics, 67 (3), 1958, p. 485–496.

[Weh98]   F. Wehrung, *The dimension monoid of a lattice* Algebra Universalis, 40 (3), 1998, p. 247–411. Or: http://www.math.unicaen.fr/~wehrung.

[WieEtAl05] K. I. Beidar, L. Márki, R. Mlitz, R. Wiegandt, *Primitive Involution Rings* Acta Math. Hungar., 109 (4), 2005, p. 357–368.

Personal Details

Name	Niklas Benjamin Niemann
Date of Birth	23/04/1977
Place of Birth	Frankfurt am Main, Germany
Country of Citizenship	Federal Republic of Germany

Education

1/12/2003 – date	PhD Candidate TU Darmstadt
25/11/2003	Diploma in Mathematics TU Darmstadt
10/2001 – 11/2003	Main Studies in Mathematics TU Darmstadt
10/2000 – 6/2001	Visiting Student (Erasmus-Program) Trinity College Dublin, Ireland
10/1997 – 9/2000	Elementary Studies in Mathematics TU Darmstadt
10/1996 – 9/1997	Civilian Service
19/06/1996	Allgemeine Hochschulreife Heinrich-Mann-Schule Dietzenbach

Scholarships

5/2005 – 3/2007	PhD Scholarship Studienstiftung des Deutschen Volkes
11/2004 – 4/2005	PhD Scholarship TU Darmstadt

Darmstadt, April 2007   Niklas Niemann